CHEF PATRICK MOULD

RECIPES FROM A CHEF

PHOTOGRAPHS BY

DEBBIE FLEMING CAFFERY

SPONSORED BY CHURCH POINT WHOLESALE'S

FROM THE KLFY TV-10 SERIES "THE CHEF"

This Book is dedicated to:

My Son Ethan: The next generation of Foodie.
My Step-son Colby: The best drummer I know.

The memory of Dr. Tommy Comeaux-
Physician-Musician-Friend
Yo, Tommy!

To the people of Louisiana;
whose joie de vivre, cultural diversity and love of food
has created the greatest regional cuisine in the world.

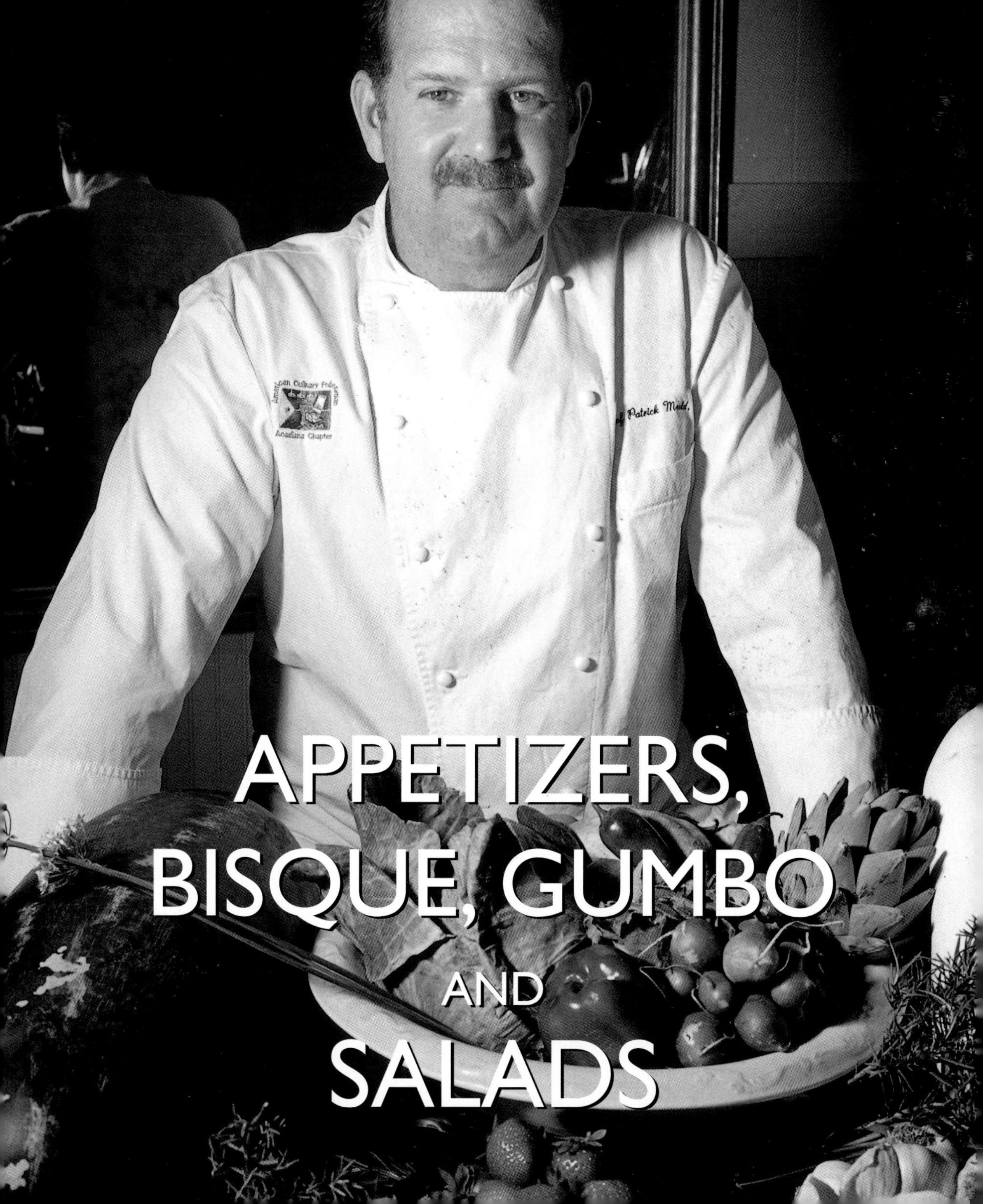

APPETIZERS, BISQUE, GUMBO AND SALADS

APPETIZERS, BISQUE, GUMBO AND SALADS

APPETIZERS

Coconut-Battered Shrimp with Orange Marmalade Dipping Sauce
Crawfish and Corn Beignets
Italian Chicken Ribbons with Plum Jam Dipping Sauce
Grilled Chicken Quesadillas
Louisiana Shrimp Dip
Mexicali Rotel Tomato-Cheese Dip
Shrimp and Mushroom Dip
Pork and Shrimp Patties with Rémoulade Sauce
Smokehouse Cocktail Meatballs

BISQUE & GUMBO

Broccoli and Tasso Cheese Bisque
Louisiana Corn and Crab Bisque
Duck, Fresh Sausage and Mirliton Gumbo
Chicken and Three Mushroom Bisque
Crawfish and Artichoke Bisque
Smoked Cornish Game Hen and Sausage Gumbo

SALADS

Broiled Portobello Mushroom Salad
Garlic Grilled Chicken with Herb Vinaigrette and Greens
Herb Vinaigrette Dressing
Marinated Shrimp and Artichoke Salad
Spicy Cajun Dressing

COCONUT-BATTERED SHRIMP WITH ORANGE MARMALADE DIPPING SAUCE

If you're a fan of sweet and spicy dishes like I am, this recipe is for you. If you like dishes on the spicier side, you can add more horseradish and hot sauce.

- *3 cups cooking oil*
- *1 ¼ cups flour*
- *3 teaspoons Tony Chachere's Creole Seasoning, (divided)*
- *2 eggs*
- *½ cup milk*
- *1 cup shredded coconut*
- *1 pound 15-20 count shrimp*, peeled and deveined*

ORANGE MARMALADE DIPPING SAUCE

- *½ cup orange marmalade*
- *1 tablespoon horseradish*
- *2 teaspoons Dijon mustard*
- *2 teaspoons lemon juice*
- *¼ teaspoon hot sauce*
- *¼ teaspoon salt*

DIRECTIONS

1. Heat cooking oil to 325°.
2. In a bowl stir together 1 cup of flour with 1 teaspoon Creole Seasoning and set aside. In another bowl place eggs, milk and 1 teaspoon Creole Seasoning. Whip together.
3. In another bowl stir together remaining ¼ cup flour and shredded coconut.
4. Place shrimp in another bowl and season with remaining teaspoon of Creole Seasoning.
5. Dip each shrimp first in flour, then in milk-egg wash, then roll in coconut-flour mixture until coated. Repeat until all shrimp are battered.
6. Drop shrimp into cooking oil and fry until golden brown and floating.
7. Serve with Orange Marmalade Dipping Sauce.

ORANGE MARMALADE DIPPING SAUCE

Combine all ingredients in a bowl and mix until thoroughly blended.

Yields 12 to 15 shrimp.

* *15-20 count denotes the number of shrimp per pound; the larger the number, the smaller the shrimp.*

CRAWFISH AND CORN BEIGNETS

When most people think of beignets, they think of the French Quarter powdered sugar variety. These are savory little clouds of crawfish and corn that make an excellent first course to any dinner. The Spicy Cajun Dressing on page 40 is a great dipping sauce to accompany this dish.

- 4 *cups cooking oil*
- 3 *cups flour*
- 2 *cups milk*
- 1 *tablespoon baking powder*
- 1 *tablespoon Tony Chachere's Creole Seasoning*
- 1 *tablespoon minced garlic*
- 1 *teaspoon dried thyme leaves*
- 1 *teaspoon hot sauce*
- 1 *pound crawfish tails, coarsely chopped*
- 1 *(16-ounce) can whole kernel corn, drained*
- ¼ *cup minced parsley*
- ¼ *cup chopped green onions*

DIRECTIONS

1. Heat oil until it reaches 350°.
2. In a large bowl blend together flour, milk, baking powder, Tony Chachere's Creole Seasoning, garlic, thyme and hot sauce. Stir until batter is formed.
3. Stir in remaining ingredients until all are incorporated.
4. Drop batter by the spoonful into hot grease, being careful not to splash.
5. Cook beignets for 5 minutes after they float to the top of pot, flipping occasionally.
6. Serve with spicy Cajun Dressing on page 40.

Yields 2 dozen Beignets.

ITALIAN CHICKEN RIBBONS WITH PLUM JAM DIPPING SAUCE

This is my version of a recipe from my early days of working as a cook for Michael Doumit at Michael's Catering, which later went on to become à la Carte Catering. Dickie Torres, the talented head of the kitchen at the time, was an early inspiration to me.

4 cups cooking oil
2 pounds boneless chicken breast strips
1 tablespoon lemon juice
1 teaspoon dried basil leaves
1 teaspoon dried oregano leaves
1 teaspoon onion powder
1 teaspoon garlic powder
1 teaspoon Tony Chachere's Creole Seasoning
¼ teaspoon salt
1 cup flour
1 cup Italian bread crumbs
1 cup shredded Parmesan cheese
1 egg
¼ cup buttermilk
¼ cup milk

PLUM JAM DIPPING SAUCE

¾ cup plum jam
2 tablespoons lemon juice
1 tablespoon horseradish
1¼ teaspoons hot sauce
1 teaspoon Tony Chachere's Creole Seasoning
1 teaspoon Worcestershire sauce
½ teaspoon garlic powder

DIRECTIONS

1. Heat cooking oil to 350°.
2. In a bowl stir together chicken breast strips, lemon juice, basil, oregano, onion powder, garlic powder, Creole seasoning and salt. Mix thoroughly.
3. Place flour in a bowl. In another bowl, stir together Italian bread crumbs and Parmesan cheese. In a third bowl, whip together egg, buttermilk and milk.
4. Start battering process. First coat chicken strips with flour, then dip in milk-egg batter, then roll in bread crumbs until coated.
5. Carefully drop breaded chicken strips into heated oil several at a time and cook until strips start to float and turn crispy. Repeat process until all chicken strips are cooked.
6. To make Plum Jam Dipping Sauce, combine all ingredients in a mixing bowl and mix thoroughly.
7. Serve chicken strips with Plum Jam Dipping Sauce.

Yields 10 appetizer portions.

GRILLED CHICKEN QUESADILLAS

The avocado dressing is what really makes this dish. The thinner you slice the chicken breast, the easier the quesadillas will be to eat.

4 (8-ounce) boneless chicken breasts
4 tablespoons olive oil
4 teaspoons chili powder
2 teaspoons cumin
2 teaspoons salt
12 corn tortillas
¾ cup shredded Monterey Jack cheese
¾ cup shredded cheddar cheese
¾ cup salsa
6 tablespoons olive oil, divided

AVOCADO-SOUR CREAM DRESSING

1 avocado, peeled and mashed
3 tablespoons sour cream
1 tablespoon lime juice
1 tablespoon minced jalapeño pepper
1 tablespoon minced fresh cilantro
½ teaspoon Tony Chachere's Creole Seasoning

DIRECTIONS

1. Heat outdoor grill or skillet on stove. Season chicken breasts with olive oil, chili powder, cumin and salt. Cook chicken breasts; cut each breast into thin strips and set aside. You should get about 9 strips per breast.
2. Combine all ingredients for avocado-sour cream dressing in a small bowl and blend together. Divide among 6 tortillas and spread evenly. Combine cheeses and sprinkle 2 tablespoons per tortilla. Place 6 chicken strips in a spiral over cheeses. Sprinkle additional 2 tablespoons of cheese over chicken.
3. Divide and spread salsa over 6 remaining tortillas. Place onto chicken tortilla (salsa side down) and press down, forming a sandwich.
4. Heat 1 tablespoon olive oil in a nonstick sauté pan. Brown one side of quesadilla, about 1 to 2 minutes. Very carefully flip over and continue to cook until other side is browned and cheese is melted.
5. Allow to cool slightly, then cut into quarters. Cook remaining quesadillas in the same fashion.

Yields 24 Quesadilla Quarters.

LOUISIANA SHRIMP DIP

Making Creole cream cheese from fresh cow milk was once a weekly ritual in Southwestern Louisiana. Today we substitute the store-bought version. I've combined the cream cheese with shrimp and spices to make an excellent party dip.

2 cups water
1 tablespoon plus 1 teaspoon Tony Chachere's Creole Seasoning, divided
1 pound small shrimp, peeled
1 pound cream cheese, at room temperature
5 tablespoons reserved shrimp broth
2 tablespoons lemon juice
2 teaspoons onion powder
2 teaspoons hot sauce
1 teaspoon garlic powder
½ teaspoon salt
½ teaspoon white pepper
2 tablespoons minced green onion
2 tablespoons minced parsley

DIRECTIONS

1. Bring 2 cups of water to boil in medium pot. Add 1 tablespoon of Creole Seasoning.
2. Add shrimp, bring back to boil and simmer for 2 minutes. Strain and reserve 5 tablespoons shrimp broth. Chill shrimp in refrigerator. This can be done a day ahead of time.
3. Place cream cheese, reserved shrimp broth and lemon juice in bowl. Beat mixture until smooth.
4. Add onion powder, hot sauce, garlic powder, salt and white pepper and continue beating mixture until completely blended.
5. Coarsely chop shrimp then fold into cream cheese mixture cooked shrimp, green onions and parsley.
6. Serve with chips and crackers.

Yields 2 pints of dip.

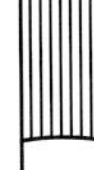

MEXICALI ROTEL TOMATO-CHEESE DIP

The combination of the chili powder and the cumin gives this dip a unique southwestern flavor.

- 1 *pound ground chuck*
- 1 *cup minced onions*
- 2 *teaspoons chili powder*
- 1 *teaspoon cumin*
- 2 *cups heavy whipping cream*
- 1 *(10-ounce) can Rotel tomatoes, drained*
- 2 *tablespoons minced jalapeño peppers*
- 1 *tablespoon Tony Chachere's Creole Seasoning*
- 1 *tablespoon minced garlic*
- 2 *cups shredded cheddar cheese*
- 2 *tablespoons minced parsley*

DIRECTIONS

1. In medium skillet over medium heat, cook ground chuck, onion, chili powder and cumin until browned. Drain off excess fat.
2. Add heavy whipping cream, Rotel tomatoes, jalapeño peppers, Creole seasoning and garlic. Cook until cream is reduced by half and slightly thickened.
3. Lower fire to simmer and stir in shredded cheddar. Continue to cook until cheese is melted and dip is creamy. Stir in parsley.
4. Dip is best served warm with tortilla chips.

Yields approximately 4 cups of dip.

SHRIMP AND MUSHROOM DIP

You can use your favorite variety of mushroom for this dip; however, avoid canned mushrooms at all cost. There are two ingredients in the world of food that I unequivocally refuse to use; frozen crabmeat and canned mushrooms.

- *1 stick unsalted butter*
- *2 cups sliced mushrooms*
- *1 cup chopped onion*
- *½ cup chopped green bell pepper*
- *½ cup chopped red bell pepper*
- *½ cup chopped celery*
- *1 tablespoon minced garlic*
- *2 teaspoons Tony Chachere's Creole Seasoning*
- *4 tablespoons flour*
- *2 cups heavy cream*
- *1 cup shredded cheddar cheese*
- *1 teaspoon hot sauce*
- *⅛ teaspoon white pepper*
- *1 pound medium shrimp, peeled and deveined*
- *3 tablespoons minced parsley*

DIRECTIONS

1. Heat butter in medium saucepot. Add mushrooms, onions, bell peppers, celery and garlic. Cook for 5 minutes, stirring occasionally.
2. Add Creole seasoning and flour and cook for 3 minutes, being careful not to brown the flour.
3. Add heavy cream, cheese, hot sauce and white pepper. Bring to a boil. Lower fire and simmer for 10 minutes, stirring constantly to make sure sauce does not stick.
4. Add shrimp and continue to simmer, stirring occasionally for additional 10 minutes.
5. Add parsley. Serve with crackers.

Yields approximately 6 cups of dip.

PORK AND SHRIMP PATTIES WITH RÉMOULADE SAUCE

This recipe is my interpretation of a dish I had at a well-known neighborhood restaurant in New Orleans, Uglesich's. If you go, be sure to get there early — they don't have many tables, and they fill up fast.

- 1 *pound* **Bryan Hot Pampered Pork Sausage**
- 1 *cup chopped onions*
- ½ *cup chopped celery*
- ½ *cup chopped green bell pepper*
- 1 *tablespoon minced garlic*
- 1 *tablespoon plus 2 teaspoons Tony Chachere's Creole Seasoning, divided*
- 1 *teaspoon hot sauce*
- ½ *teaspoon dried thyme leaves*
- 4 *tablespoons flour*
- 1 *pound small shrimp, peeled*
- 1 *cup beef broth*
- ½ *cup grated cheddar cheese*
- ½ *cup chopped parsley*
- ¼ *cup chopped green onions*
- 2 *eggs*
- 1 *cup milk*
- 2 *cups flour*
- 2 *cups corn meal*
- 4 *tablespoons olive oil*

RÉMOULADE SAUCE

- 2 *cups mayonnaise*
- 2 *tablespoons Creole mustard*
- 2 *tablespoons lemon juice*
- 2 *tablespoons Tiger Sauce (optional)*
- 2 *tablespoons white wine*
- 2 *tablespoons minced green onions*
- 2 *tablespoons minced parsley*
- 1 *tablespoon horseradish*
- 1 *tablespoon paprika*
- 1 *teaspoon Tony Chachere's Creole Seasoning*
- 1 *teaspoon hot sauce*
- ½ *teaspoon Worcestershire sauce*
- ½ *teaspoon minced garlic*

DIRECTIONS

1. Heat saucepot over medium heat.
2. Add **Bryan's Hot Pampered Pork Sausage** and cook for 10 minutes, stirring occasionally until brown. Add onion, celery, bell pepper and cook an additional 10 minutes, continuing to stir.
3. Add garlic, 1 tablespoon Creole seasoning, hot sauce and thyme, and continue cooking another five minutes.

4. Stir in flour and cook another minute. Add shrimp and cook for 5 more minutes.
5. Add beef broth and cook until slightly thickened.
6. Stir in cheese, parsley and green onions. Reduce heat and cook until cheese melts.
7. Pour into bowl and place in refrigerator. Chill for 3 hours until firm.
8. In a bowl beat together eggs and milk; set aside. In another bowl, stir together flour and 1 teaspoon of Creole seasoning. In another bowl, stir together corn meal and remaining Creole seasoning,
9. Scoop ⅓ cup of shrimp and pork mixture into flour, then milk-egg batter and roll in yellow corn meal. Flatten into a patty.
10. Heat 2 tablespoons olive oil in a medium sauté pan over medium heat. Place 6 battered patties in pan and sauté until brown on one side, flip and continue to cook until browned. Repeat process with remaining 6 patties.
11. Top each patty with a dollop of Rémoulade Sauce before serving.

RÉMOULADE SAUCE

1. Combine all ingredients in a bowl and beat until blended.
2. Refrigerate until chilled.

Yields 12 appetizer servings.

SMOKEHOUSE COCKTAIL MEATBALLS

This recipe will be a surefire hit at your next party. The cane syrup and hot sauce mix well with your favorite BBQ sauce to give the sauce a nice tang.

COCKTAIL MEATBALLS

- 1 *pound ground beef*
- 3 *tablespoons minced onion*
- 2 *tablespoons minced bell pepper*
- 2 *tablespoons minced celery*
- 1 *tablespoon minced garlic*
- 1 *tablespoon minced parsley*
- 1 *tablespoon minced green onion*
- 1 *tablespoon hot sauce*
- 1 *teaspoon Tony Chachere's Creole Seasoning*
- ¼ *teaspoon salt*
- 2 *eggs*
- 1 *cup Italian bread crumbs*
- 1 *tablespoon vegetable oil*

SAUCE

- ¾ *cup your favorite BBQ Sauce*
- ¼ *teaspoon Tony Chachere's Creole Seasoning*
- 1 *tablespoon Worcestershire sauce*
- 2 *tablespoons cane syrup*
- ¼ *tablespoon hot sauce*
- ½ *cup beef broth*

DIRECTIONS

1. Combine all ingredients for cocktail meatballs in a bowl and mix thoroughly.
2. Form into approximately 20 cocktail size meatballs.
3. Heat oil in large skillet over medium heat.
4. Add meatballs and brown slightly on all sides. Add remaining ingredients for sauce and simmer until meatballs are cooked and sauce starts to thicken slightly, about 15 minutes.
5. Serve as an appetizer or over pasta as a main course.

Yields 20 meatballs.

BROCCOLI AND TASSO CHEESE BISQUE

Broccoli and cheese are a match made in heaven. Be careful not to overcook the broccoli in order to maintain its green color and crunchiness.

- ½ *cup cooking oil*
- 1 *pound tasso, medium coarse chopped*
- 1 *cup chopped onion*
- ¾ *cup chopped red bell pepper*
- ½ *cup chopped green bell pepper*
- ½ *cup chopped celery*
- 1 *tablespoon minced garlic*
- 1 *cup flour*
- 8 *cups chicken broth, heated*
- 2 *cups heavy whipping cream*
- 1 *pound cheddar cheese*
- 1 *pound fresh broccoli, cut into pieces*
- 2 *teaspoons Tony Chachere's Creole Seasoning*
- ½ *teaspoon white pepper*
- ¼ *cup minced parsley*
- ¼ *cup chopped green onions*

DIRECTIONS

1. Heat cooking oil in a large saucepot over medium heat. Add tasso and cook for 5 minutes. Be careful not to brown it.
2. Add onion, bell peppers, celery and garlic. Cook for 5 minutes.
3. Add flour and stir until a roux is formed. Do not brown flour. Add the heated chicken broth and whip until a light sauce is formed. Simmer for 5 minutes.
4. Stir in heavy cream, cheddar cheese, broccoli, Creole seasoning and white pepper. Cover and simmer for 10 minutes, stirring occasionally.
5. Add parsley and green onions.
6. Divide into 6 large bowls and serve.

Yields 6 servings.

Road Sign—Delcambre Highway

LOUISIANA CORN AND CRAB BISQUE

This is the quintessential Louisiana Bisque. You can use any fresh seafood, but jumbo lump crab, in my opinion, is the best. The flavor obtained when it is simmered in the cream is "to die for"... worry about the calories later.

- 2 *tablespoons unsalted butter*
- 1 *cup chopped onion*
- ½ *cup chopped green bell pepper*
- ½ *cup chopped celery*
- ¼ *cup chopped red bell pepper*
- 1 *tablespoon minced garlic*
- 2 *cups chicken broth*
- ½ *cup dry white wine*
- ¾ *teaspoon dried thyme leaves*
- ½ *cup blond roux (¼ cup vegetable oil and ¼ cup flour)*
- 3½ *cups heavy whipping cream*
- 1 *teaspoon salt*
- 1 *teaspoon hot sauce*
- 1 *cup cooked corn*
- 1 *pound lump crabmeat*
- 1 *tablespoon chopped parsley*
- 1 *tablespoon chopped green onion*
- 16 *crab claws, optional*

DIRECTIONS

1. Heat the butter over a low to medium heat in a 4-quart saucepot.
2. Add onion, green bell pepper, celery, red pepper and garlic and cook for 1 minute.
3. Add chicken broth, white wine and thyme. Bring to boil.
4. In a small bowl make blond roux by combining oil and flour and stirring until a smooth paste is formed.
5. Whip roux into broth mixture until mixture begins to thicken. Whip in cream, reduce heat to a simmer and continue to cook until cream is blended in and beginning to thicken.
6. Add salt, hot sauce and corn. Simmer 5 minutes.
7. Being very careful not to break up lumps, stir in crabmeat, parsley and green onions. Simmer until heated.
8. Divide into 4 large bowls. Garnish with crab claws, if desired.

Yields 4 servings.

DUCK, FRESH SAUSAGE AND MIRLITON GUMBO

This is a hunter's dream. If you don't have a hunter in the family, a domestic duck works fine. The mirlitons, also referred to as vegetable pears or chayote squash, make a nice addition to the gumbo.

- *1 (5 to 6-pound) domestic duckling*, cut up 8 ways*
- *2 tablespoons Worcestershire sauce*
- *4 teaspoons Tony Chachere's Creole Seasoning, divided*
- *1½ teaspoons hot sauce, divided*
- *1 teaspoon granulated garlic*
- *1 teaspoon granulated onion*
- *2 quarts chicken broth**
- *1 cup chopped onion, divided*
- *½ cup chopped celery, divided*
- *½ cup chopped bell pepper, divided*
- *½ cup* ***Savoie's Dark Roux***
- *3 bay leaves*
- *1 pound fresh pork sausage*
- *2 large mirlitons, peeled, seeded and cut into medium cubes*
- *¼ teaspoon salt*
- *¼ cup sliced green onions*
- *8 cups cooked Toro Rice*

DIRECTIONS

1. Season duckling with Worcestershire sauce, 3 teaspoons of Creole seasoning, 1 teaspoon hot sauce, granulated garlic and granulated onion. Marinate overnight in refrigerator.

2. Place duck on baking pan and roast in a 400° oven for 45 minutes. Remove duck from pan and discard fat.

3. In a large saucepot, combine chicken broth with ½ cup onion, ¼ cup celery and ¼ cup bell pepper. Add **Savoie's Dark Roux,** remaining hot sauce, bay leaf, unsliced fresh sausage and roasted duckling. Bring to boil, lower fire, cover pot and simmer for 30 minutes*.

4. Remove cooked sausage; cool and cut into slices. Set aside. Add remaining onion, celery, bell pepper and garlic. Cover and simmer for an additional 30 minutes.

5. Add sausage, mirliton and salt. Continue to simmer uncovered for 15 minutes. Cover and cook for 15 more minutes. Stir in green onions and serve over rice.

Yields 6 to 8 servings.

** If you are using wild ducks, cook the ducks longer to tenderize. The length of time will depend upon the toughness of the ducks. You will also have to increase the chicken broth by ½ quart.*

Sportsman's Paradise

CHICKEN AND THREE MUSHROOM BISQUE

A lot of recipes from South Louisiana call for cream of mushroom soup. I have also added cream of chicken soup and a sauté of wild mushrooms to this recipe for enhanced flavor.

1 stick unsalted butter
1 pound boneless chicken breast, cubed
2 teaspoons Tony Chachere's Creole Seasoning, divided
¾ cup chopped onions
½ cup chopped red bell pepper
¼ cup chopped green bell pepper
½ cup chopped celery
1 tablespoon minced garlic
1¼ teaspoons dried tarragon leaves
6 cups sliced assorted fresh mushrooms: shiitake, white and Italian
½ cup flour
2 cups chicken broth
1 (6-ounce) can ***Campbell's Cream of Chicken Soup***
1 (6-ounce) can ***Campbell's Cream of Mushroom Soup***
2 cups half & half cream
1 tablespoon hot sauce
½ cup chopped green onions
¼ cup chopped parsley

DIRECTIONS

1. Heat butter over medium heat in large saucepot, being careful not to burn. Season chicken with 1 teaspoon of Creole seasoning, add to pot and cook until chicken starts to brown slightly, about 8 to 10 minutes.
2. Add onion, bell peppers and celery and cook for an additional two minutes. Add garlic and tarragon and cook for an additional minute. Then add mushrooms and cook for 5 minutes.
3. Stir in flour and cook for 1 minute until flour has been absorbed and a roux begins to develop. Do not brown flour. Stir in chicken broth, **Campbell's Cream of Chicken Soup** and **Campbell's Cream of Mushroom Soup** and stir constantly until bisque begins to thicken.
4. Reduce heat to medium and stir in half & half and hot sauce until blended. Cover pot and simmer bisque for 10 minutes, stirring occasionally.
5. Stir in green onions and parsley.

Yields 8 servings.

CRAWFISH AND ARTICHOKE BISQUE

Serve up this hearty bisque with a crusty loaf of French bread and a bottle of your favorite wine and you can't go wrong. Make sure to include a loved one.

- 1 *stick unsalted butter*
- 1 *cup chopped onion*
- ½ *cup chopped celery*
- ½ *cup chopped green bell pepper*
- ¼ *cup chopped red bell pepper*
- 1 *teaspoon chopped garlic*
- 1 *teaspoon dried thyme leaves*
- 2 *bay leaves*
- 2 *(14-ounce) cans quartered artichoke hearts, drained*
- ½ *cup flour*
- 2 *cups chicken broth*
- 3 *cups half & half cream*
- 1 *pound crawfish tails*
- 1 *teaspoon hot sauce*
- ½ *cup chopped green onions*
- ¼ *cup minced parsley*

DIRECTIONS

1. Heat butter in a large saucepot over medium heat. Add onion, celery, bell peppers, garlic, thyme and bay leaves, cook for 5 minutes.
2. Add artichoke hearts and cook an additional 2 minutes. Add flour and stir until incorporated.
3. Stir in chicken broth and cook for 2 minutes until chicken broth starts to thicken. Stir in half & half cream and simmer for 2 minutes until bisque is smooth and creamy.
4. Add crawfish tails and hot sauce and simmer for 5 minutes.
5. Stir in green onions and parsley.

Yields 6 to 8 servings.

SMOKED CORNISH GAME HEN AND SAUSAGE GUMBO

Smoking meats is one of the greatest pleasures known to a food lover, and it's not as difficult as you might imagine. You can intensify the smoky flavor of any dish by making a stock out of the bones of the meat you have smoked, as I have done in this dish.

- *2 (20-ounce) Cornish game hens*
- *3 tablespoons Tony Chachere's Creole Seasoning, divided*
- *¾ gallon smoked hen stock**
- *1 cup* ***Savoie's Dark Roux***
- *1 pound smoked sausage, sliced, browned and drained*
- *1 cup chopped onion*
- *½ cup chopped celery*
- *½ cup chopped green bell pepper*
- *1 tablespoon minced garlic*
- *1 tablespoon Worcestershire sauce*
- *2 teaspoons hot sauce*
- *1 teaspoon salt*
- *½ cup chopped green onions*
- *¼ cup minced parsley*
- *4 cups cooked Toro Rice*

DIRECTIONS

1. Season Cornish hens with 1 tablespoon of Creole seasoning and smoke in a conventional smoker. Allow hens to cool. Debone discarding skin and reserve bones for stock. Dice meat and set aside.
2. Place smoked hen stock* in large saucepot and bring to boil. Add **Savoie's Dark Roux,** lower heat and simmer 30 minutes.
3. Add smoked sausage, onion, celery, bell pepper, garlic, remaining 2 tablespoons Creole seasoning, Worcestershire, hot sauce and salt. Simmer for an additional 30 minutes.
4. Add diced Cornish hen meat and simmer for 15 minutes.
5. Stir in green onions and parsley. Divide Toro Rice into 8 large bowls and ladle in gumbo.

Yields 8 servings.

**Smoked hen stock: Cover hen bones with water. Add end pieces from onion, celery and bell pepper, and bring to boil. Lower fire and simmer for 1½ hours. Strain out solids and reserve stock. This can be done the day prior to cooking the gumbo. After stock is chilled, skim off the fat that rises to the top.*

Paying homage to the chicken-Mamou Mardi Gras

BROILED PORTOBELLO MUSHROOM SALAD

Portobello mushrooms are known for their meaty texture and, although I am broiling them, they are even better grilled.

- ⅛ *cup balsamic vinegar*
- 1 *teaspoon dried tarragon leaves*
- 1 *teaspoon minced garlic*
- 1 *teaspoon Tony Chachere's Creole Seasoning*
- ¼ *teaspoon hot sauce*
- ½ *cup extra virgin olive oil, divided*
- 4 *portobello mushrooms caps, stems removed*
- ½ *cup water*
- 2 *tablespoons Dijon mustard*
- ¼ *teaspoon salt*
- *Assorted salad greens for two large dinner salads*

DIRECTIONS

1. Preheat oven broiler.
2. In a bowl, combine balsamic vinegar, tarragon, garlic, Creole seasoning and hot sauce, and whip together.
3. Slowly add ¼ cup olive oil until completely incorporated. Add mushrooms and marinate for 30 minutes at room temperature.
4. Place marinade and mushrooms bottom side up in a baking dish and broil in oven for five minutes. Flip mushrooms and broil for an additional 2 minutes.
5. Remove mushrooms from oven, place on a plate and keep warm. Pour water into hot baking dish, scrape juices from bottom and pour reserved liquid in a bowl. Allow to cool slightly.
6. Whip mustard and salt into reserved liquid; then slowly whip in remaining olive oil and chill.
7. Slice broiled mushroom caps and place on 2 plates that have been lined with assorted greens. Top with dressing.

Yields 2 entrée salads or 4 side salads.

GARLIC GRILLED CHICKEN WITH HERB VINAIGRETTE AND GREENS

This is a great healthy summer salad. You can serve the chicken warm or chilled — I prefer it warm. Serve with a loaf of crusty French bread and a chilled Chardonnay.

MARINADE

- *4 tablespoons minced garlic*
- *3 tablespoons red wine vinegar*
- *2 tablespoons hot sauce*
- *4 teaspoons Tony Chachere's Creole Seasoning*
- *½ cup olive oil*
- *8 boneless skinless chicken breasts*

DRESSING

- *4 teaspoons white wine vinegar*
- *3 teaspoons Dijon mustard*
- *1½ teaspoons dried basil leaves*
- *1 teaspoon minced garlic*
- *1 teaspoon Tony Chachere's Creole Seasoning*
- *1 teaspoon hot sauce*
- *½ teaspoon dried thyme leaves*
- *¼ teaspoon salt*
- *1 cup extra virgin olive oil*
- *Assorted greens*

DIRECTIONS

1. In a bowl, whip together garlic, red wine vinegar, hot sauce and Creole seasoning.
2. Slowly whip in olive oil until all is incorporated. Add chicken breasts and marinate a minimum of 2 hours under refrigeration.
3. Fire up pit and grill chicken, preferably over hardwood, for five minutes on each side, depending on size of chicken breasts. Chicken can also be cooked in a sauté pan on the stove.

SALAD DRESSING

1. In the bowl whip together all ingredients except extra virgin olive oil and assorted greens.
2. Slowly whip in olive oil until all is incorporated.
3. Slice cooked chicken breasts into strips and place on top of assorted greens. Drizzle each salad with dressing.

Yields 8 servings.

HERB VINAIGRETTE DRESSING

Use only the finest extra virgin olive oil you can find for this salad dressing.

- *3 tablespoons balsamic vinegar*
- *2 tablespoons Dijon mustard*
- *1 tablespoon minced garlic*
- *1½ teaspoons dried basil leaves*
- *1 teaspoon dried oregano leaves*
- *1 teaspoon Tony Chacere's Creole Seasoning*
- *½ teaspoon dried thyme leaves*
- *¼ teaspoon black pepper*
- *¼ teaspoon salt*
- *¾ cup extra virgin olive oil*

DIRECTIONS

1. In a bowl, whip together all ingredients except olive oil until blended.

2. Slowly pour in olive oil while whipping constantly, until all of the oil is incorporated.

Yields approximately 1 cup of dressing.

MARINATED SHRIMP AND ARTICHOKE SALAD

Be sure not to overcook your shrimp.

- 2 *cups water*
- 1 *tablespoon plus 1 teaspoon Tony Chachere's Creole Seasoning, divided*
- 1 *pound medium-sized shrimp, peeled*
- 3 *tablespoons balsamic vinegar*
- 2 *tablespoons Dijon mustard*
- 1 *tablespoon minced garlic*
- 1½ *teaspoons dried basil leaves*
- 1 *teaspoon dried oregano leaves*
- ½ *teaspoon dried thyme leaves*
- ¼ *teaspoon black pepper*
- ¼ *teaspoon salt*
- ¾ *cup extra virgin olive oil*
- 2 *(14-ounce) cans quartered artichoke hearts, drained*
- *Assorted greens*
- *Tomato garnish*
- *Goat cheese garnish*

DIRECTIONS

1. Bring 2 cups of water to boil in medium pot. Add 1 tablespoon of Creole seasoning.
2. Add shrimp, bring back to boil and simmer for 2 minutes. Strain shrimp and chill in refrigerator.
3. In a medium bowl, whip together balsamic vinegar, mustard, garlic, remaining Tony Chachere's Creole Seasoning, basil, oregano, thyme, black pepper and salt until blended.
4. Slowly pour in olive oil while whipping constantly until all of the oil is incorporated.
5. Stir in chilled shrimp and quartered artichoke hearts. Marinate for 2 hours under refrigeration. Serve marinated shrimp and artichokes over bed of greens. Garnish with tomatoes and goat cheese.

Yields 8 servings.

SPICY CAJUN DRESSING

This recipe is not only a great salad dressing but can also be used as a dipping sauce.

- *4 cups mayonnaise*
- *½ cup minced parsley*
- *½ cup minced green onions*
- *¼ cup minced celery*
- *¼ cup Dijon mustard*
- *6 tablespoons lemon juice*
- *6 tablespoons ketchup*
- *4 tablespoons minced capers*
- *2 tablespoons Tony Chachere's Creole Seasoning*
- *1 tablespoon hot sauce*
- *1 tablespoon Worcestershire sauce*

DIRECTIONS

1. Combine all ingredients in a small mixing bowl and whip until blended. Or place in bowl of a food processor and process on high speed for 2 minutes until blended.

2. Chill before serving. May be used as a salad dressing or as a dipping sauce.

Yields approximately 6 cups of dressing.

SEAFOOD

SEAFOOD

Alligator and Smoked Andouille Sauce Piquante
Barbecue Skewers of Shrimp
Broiled Catfish with Lump Crabmeat Sauté
Bronzed Catfish Fillets
Cajun Seafood Fry with Tartar Sauce
Cane Syrup Glazed Prawns with Lemon-Garlic Butter Sauce
Catfish Courtbouillion
Catfish Louisiana
Crawfish Corn Macque Choux
Crawfish Dumpling Étouffée
Crawfish-Mushroom Casserole
Fried Frog Legs and Fettuccine Alfredo
Grilled Soft-Shell Crabs with Lump Crabmeat Sauté
Louisiana Blue Point Lump Crabmeat Alfredo
Louisiana Crawfish au Gratin
Marinated Grilled Catfish
Oriental Fish Fry
Pan Sautéed Flounder with Oyster Meunière Sauce
Pat's Crawfish Étouffée
Pecan-Crusted Catfish, Topped with Spicy Lemon-Beurre Blanc
Seafood and Tasso Pasta
Shrimp, Mirliton and Corn Macque Choux
Shrimp and Egg Stew
Shrimp and Tasso Jambalaya
Shrimp and Tasso Sauté
Shrimp Creole
Southwestern Barbecue Shrimp and Pasta

BARBECUE SKEWERS OF SHRIMP

Jack Miller's Barbecue Sauce is a true Louisiana product. There are few backyard cookouts in Cajun Country that don't utilize this excellent sauce for basting foods that are cooking on the pit!

- *1 pound 15 to 20 count shrimp*, peeled and deveined*
- *2 teaspoons **Jack Miller's Barbecue Seasoning***
- *1 tablespoon lemon juice*
- *2 tablespoons olive oil, divided*
- *¼ cup **Jack Miller's Barbecue Sauce***

DIRECTIONS

1. Preheat gas grill or fire up the barbecue pit.
2. In a bowl toss together shrimp, **Jack Miller's Barbecue Seasoning,** lemon juice and 1 tablespoon olive oil. Marinate shrimp for 30 minutes in refrigerator. Divide shrimp and thread onto 2 wooden skewers.
3. In another bowl, stir together **Jack Miller's Barbecue Sauce** and remaining olive oil.
4. Place skewers of shrimp on grill and baste with half of the barbecue sauce and olive oil mixture. Cover and cook for 5 minutes. Turn over and baste with remaining barbecue sauce for an additional 3 minutes.
5. Additional **Jack Miller's Barbecue Sauce** may be used for dipping shrimp.

Yields 2 servings.

**15 to 20 count denotes the number of shrimp to the pound. The smaller the number, the larger the shrimp.*

ALLIGATOR AND SMOKED ANDOUILLE SAUCE PIQUANTE

Perhaps the most recognizable creature in the Cajun Culinary Kingdom is the alligator. Every September alligator hunters wander the back roads and bayous of the Louisiana swamp in their attempt to capture these prehistoric reptiles whose meat has become a delicacy.

- *2 pounds alligator meat, cubed*
- *5 cups water*
- *2 tablespoons Tony Chachere's Creole Seasoning, divided*
- *2 tablespoons olive oil*
- *1 pound smoked andouille, sliced*
- *1 cup chopped onion*
- *½ cup chopped green bell pepper*
- *½ cup chopped celery*
- *1 tablespoon minced garlic*
- *1 (14-ounce) can diced tomatoes, drained*
- *1 (8-ounce) can tomato sauce*
- *1 (6-ounce) can tomato paste*
- *1 cup ketchup*
- *½ cup red wine*
- *3 tablespoons Savoie's Dark Roux*
- *1 tablespoon hot sauce*
- *2¼ cups reserved alligator stock*
- *1 teaspoon dried thyme leaves*
- *1 whole lemon, sliced*
- *2 bay leaves*
- *½ pound sliced mushrooms*
- *½ cup minced green onions*
- *¼ cup minced parsley*
- *10 cups cooked* ***Toro Brand Rice***

DIRECTIONS

1. In large saucepot, place alligator meat and 1 tablespoon Creole seasoning; add water and bring to boil. Lower fire and simmer for 30 minutes. Strain alligator reserving 2¼ cups liquid to use as alligator stock in recipe.
2. In large saucepot, heat oil, add andouille and cook until browned. Stir in onion, bell pepper and celery and cook for 2 minutes over high heat. Add garlic and cook for additional minute.
3. Add diced tomato, tomato sauce, tomato paste, ketchup, red wine, roux and hot sauce. Simmer over medium heat for 15 minutes, allowing water to evaporate. Be careful not to brown too much. Stir occasionally.
4. Add alligator stock, thyme, lemon, bay leaves and remaining 1 tablespoon of Creole seasoning. Lower fire, cover and simmer for 30 minutes, stirring occasionally.
5. Remove lemon slices and add sliced mushrooms and alligator meat and simmer, covered, for an additional 10 minutes.
6. Add green onions and parsley.
7. Serve each serving with one cup of cooked **Toro Brand Rice.**

Yields 10 servings.

Alligator Princess

BROILED CATFISH FILLETS WITH LUMP CRABMEAT SAUTÉ

Catfish continues to be one of this country's favorites. Be careful not to overcook the fish and use only fresh crabmeat, never frozen.

2 *(8-ounce) catfish fillets*
1 *tablespoon olive oil*
1 *tablespoon lemon juice*
¾ *teaspoon garlic powder, divided*
¾ *teaspoon onion powder, divided*
¼ *teaspoon paprika*
½ *teaspoon black pepper, divided*
¾ *teaspoon salt, divided*
2 *tablespoons unsalted butter*
1 *tablespoon lemon juice*
¼ *cup white wine*
1 *pound lump crabmeat*

DIRECTIONS

1. Preheat oven broiler. In a bowl toss together catfish, olive oil, lemon juice, ½ teaspoon garlic powder, ½ teaspoon onion powder, paprika, ¼ teaspoon black pepper and ¼ teaspoon salt. Marinate for 2 hours under refrigeration.
2. Place catfish on a greased baking pan and cook under broiler 12 to 15 minutes, until catfish is white and flaky.
3. Place butter, lemon juice, wine, and remaining salt, pepper, garlic powder and onion powder in a skillet. Simmer for 3 minutes.
4. Gently stir in the lump crabmeat, being careful not to break up the lumps; cook for 2 more minutes.
5. Top each broiled catfish with crabmeat sauce.

Yields 2 servings.

BRONZED CATFISH FILLETS

Cajun Culinary Guru Chef Paul Prudhomme first introduced the world to Louisiana cooking with his Blackened Redfish. Bronzing is an extension of the blackening process. The food is not as charred and is a little more delicate in flavor. Chefs across Louisiana owe a tremendous amount of gratitude to this Master of Cajun Cooking.

- *2 (8-ounce) catfish fillets*
- *2 tablespoons melted butter*
- *1 tablespoon lemon juice*
- *¾ teaspoon salt*
- *½ teaspoon dried basil leaves*
- *½ teaspoon dried oregano leaves*
- *½ teaspoon garlic powder*
- *¼ teaspoon dried thyme leaves*
- *¼ teaspoon black pepper*
- *¼ teaspoon onion powder*
- *¼ teaspoon paprika*
- *⅛ teaspoon cayenne pepper*
- *⅛ teaspoon white pepper*

DIRECTIONS

1. In a bowl, toss together all ingredients until blended.

2. Preheat a skillet over medium fire. Place catfish fillets in skillet, skin side up, and cook for 2 minutes. Turn over fish and cook for an additional minute. Then flip again and cook 1 more minute or until there is no resistance when you pierce the flesh with the point of a knife.*

Yields 2 servings.

**When you are able to pierce the thickest part of the fillet easily, the fish is cooked. Be extremely careful when flipping fish to avoid burning yourself. You should have your stove vent on during the cooking process to avoid filling your kitchen with smoke. If you don't have a vent you may want to cook this dish outdoors.*

CAJUN SEAFOOD FRY WITH TARTAR SAUCE

The secret to good fried seafood is the temperature in which you fry; 350° is ideal. The primary cause of greasy fried food is submerging the food being cooked into the oil before it has reached the proper temperature of 350°. Allowing the temperature *to return* to 350° is especially critical when frying multiple batches of food. A simple kitchen tool, a frying thermometer, can alleviate this problem. Another tip is to shake off any excess fish fry before you drop food into the grease. This will prevent the excess fish fry from sinking to the bottom of the pot and burning.

6 cups cooking oil
1 pound catfish, cut into chunks
1 pound large shrimp, peeled and deveined
1 pint raw oysters, drained
4 tablespoons plus 1 teaspoon Tony Chachere's Creole Seasoning, divided
3 tablespoons lemon juice
4 eggs
1 cup buttermilk
1 cup milk
6 cups fish fry

SPICY TARTAR SAUCE

3 cups mayonnaise
6 tablespoons minced capers
3 tablespoons lemon juice
2 tablespoons minced stuffed olives
2 tablespoons minced green onions
1 tablespoon minced parsley
1 tablespoon hot sauce
1 tablespoon pickle relish
1 tablespoon minced garlic
2 teaspoons red wine vinegar
1 teaspoon Tony Chachere's Creole Seasoning

DIRECTIONS

1. Heat oil to 350° on deep fat fry thermometer. In a bowl, toss together catfish, shrimp and oysters with 4 tablespoons Creole seasoning and lemon juice.
2. In a bowl, beat together eggs, buttermilk, milk and remaining Creole seasoning.
3. Dip seafood piece by piece in fish fry until coated. Dip into milk-egg batter and roll in fish fry until completely coated.
4. Fry seafood separately because they each have different cooking times. Fish take the longest. Shrimp take less time than the fish, and oysters take the least amount of time.
5. Seafood is generally cooked when it floats. Allow the seafood to cook for an additional minute longer for extra crunchiness.
6. Have extra fish fry on hand in case you need more for coating. Serve with Spicy Tartar Sauce.

SPICY TARTAR SAUCE

In a bowl, stir together all ingredients. Refrigerate until needed.

Yields 6 servings.

CANE SYRUP-GLAZED PRAWNS WITH LEMON GARLIC BUTTER SAUCE

Prawns are huge freshwater shrimp; if you are unable to locate them you may substitute the largest shrimp you can find. Make sure you have plenty of French bread on hand so you can dip into the sauce.

- 1 *pound large prawns, peeled and deveined*
- 3 *tablespoons cane syrup*
- ½ *teaspoon Tony Chachere's Creole Seasoning*
- ½ *teaspoon crushed black pepper*
- 1 *tablespoon olive oil*
- 1 *tablespoon minced garlic*
- ¼ *cup white wine*
- ¼ *cup chicken broth*
- 1 *tablespoon lemon juice*
- ½ *teaspoon dried thyme leaves*
- 4 *tablespoons chilled unsalted butter*

DIRECTIONS

1. In a bowl toss together shrimp, cane syrup, Creole seasoning and black pepper. Marinate at room temperature for 15 minutes.

2. In a skillet, heat olive oil and cook shrimp until browned. Remove shrimp from pan and keep warm. Add garlic to pan and cook until slightly brown.

3. Stir in white wine, chicken broth, lemon juice and thyme. Lower fire and simmer until reduced by half.

4. Lower heat as low as it will go and stir in chilled butter until it is melted. Remove from fire immediately. Return shrimp to pan and toss with butter sauce.

Yields 2 servings.

CATFISH COURTBOUILLION

The catfish will put off the necessary liquid you need to make the courtboullion sauce. Don't be fooled by allowing all the liquid to evaporate during the cooking process.

- *4 tablespoons **Savoie's Dark Roux***
- *1 cup minced onion*
- *½ cup minced green bell pepper*
- *½ cup minced celery*
- *1 tablespoon minced garlic*
- *1 (16-ounce) can diced tomatoes*
- *1 (8-ounce) can tomato sauce*
- *1 (6-ounce) can tomato paste*
- *1 cup chicken broth*
- *2 tablespoons lemon juice*
- *1 tablespoon Tony Chachere's Creole Seasoning*
- *1 teaspoon dried thyme leaves*
- *1 teaspoon salt*
- *1 teaspoon hot sauce*
- *1 bay leaf*
- *1 pound boneless catfish, cut into strips*
- *½ cup minced green onions*
- *¼ cup minced parsley*
- *6 cups cooked Toro Brand Rice*

DIRECTIONS

1. In large saucepot, heat roux over low heat for several minutes. Add onion, bell pepper, celery and garlic; cook for 5 minutes, stirring occasionally.
2. Add diced tomatoes, tomato sauce, tomato paste, chicken broth, lemon juice, Creole seasoning, thyme, salt, hot sauce and bay leaf. Cook uncovered until the sauce starts to thicken. Most of the water should evaporate, but be careful not to scorch the bottom.
3. Add catfish to the pot, lower heat, and cook covered for 20 minutes.
4. Being careful not to break up fish, stir green onions and parsley into the courtbouillion.
5. Serve each portion with a cup of cooked Toro Brand Rice.

Yields 6 servings.

The Guidry brothers catfishing - Delcambre canal

CATFISH LOUISIANA

This is a version of the dish that won me the Best of Show Award at the Acadiana Culinary Classic in 1984. The Acadiana Culinary Classic has been credited with starting the Renaissance of Cajun Cooking in Acadiana. I consider this distinction to be one of my greatest career achievements.

- 2 *cups cooking oil, plus 4 tablespoons, divided*
- 2 *cups flour, plus 2 tablespoons, divided*
- 1 *tablespoon plus ¼ teaspoon Tony Chachere's Creole Seasoning, divided*
- 1 *egg*
- ¼ *cup buttermilk*
- ¼ *cup milk*
- 2 *(8-ounce) catfish fillets*
- ¼ *cup chopped onion*
- ¼ *cup chopped green bell pepper*
- ¼ *cup chopped celery*
- 1 *tablespoon minced garlic*
- 1 *cup chicken broth*
- ¼ *cup white wine*
- ¼ *teaspoon paprika*
- ¼ *teaspoon hot sauce*
- ½ *pound crawfish tails**
- 1 *tablespoon chopped parsley*
- 1 *tablespoon chopped green onions*
- 2 *cups cooked* ***Toro Brand Rice***

DIRECTIONS

1. Heat cooking oil to 350°.**
2. In a bowl, blend together 2 cups flour and 1½ teaspoons Creole seasoning. In another bowl beat together egg, buttermilk and milk.
3. Toss catfish fillet in flour, then dip in egg batter, and then roll back in flour until coated. Repeat process on remaining catfish fillet.
4. Carefully drop fillet in heated oil and fry until golden brown and floating, approximately 5 minutes; cook additional minute for added crunchiness. Keep in warm oven while making sauce.
5. Heat 2 tablespoons cooking oil in saucepot. Add onion, bell pepper and celery. Cook for 1 minute. Add garlic and cook an additional minute.
6. Add broth, white wine, remaining Creole seasoning, paprika, hot sauce. In a small cup blend together remaining cooking oil and flour. Whip in until sauce thickens.***
7. Add crawfish tails, parsley and green onions. Simmer for 5 minutes.
8. Top each fried catfish fillet with sauce. Serve each portion with 1 cup cooked **Toro Brand Rice.**

Yields 2 servings.

**Crabmeat or shrimp may be substituted for crawfish.*
***If frying more than 2 portions of catfish, remember to allow frying oil to come back up to 350°.*
****If sauce thickens too much after adding white roux, thin with small amount of water.*

CRAWFISH CORN MACQUE CHOUX

"Macque Choux" is a dish that the Native Americans introduced to the Cajuns. It is best prepared with fresh corn; however you may substitute canned corn.

- ½ *stick unsalted butter*
- 3 *cups fresh corn, cut off cob*
- 1 *(10-ounce) can diced Rotel tomatoes, drained*
- 1 *cup chopped onion*
- ½ *cup chopped green bell pepper*
- ½ *cup chopped celery*
- ½ *cup chicken broth*
- 2 *cups heavy cream*
- 2 *teaspoons minced garlic*
- 2 *teaspoons Tony Chachere's Creole Seasoning*
- 1 *teaspoon salt*
- 1 *teaspoon hot sauce*
- 1 *pound crawfish tails*
- 3 *tablespoons chopped green onions*
- 2 *tablespoons minced parsley*
- 6 *cups cooked* ***Toro Brand Rice***

DIRECTIONS

1. In saucepot, heat butter until melted. Add corn and cook for 2 minutes.
2. Add Rotel tomatoes, onion, bell pepper, celery and chicken broth; turn heat up and cook until chicken broth has evaporated.
3. Stir in heavy cream, garlic, Creole seasoning, salt and hot sauce. Bring to a boil.
4. Add crawfish tails, lower heat and simmer for 10 minutes, stirring occasionally until cream reduces and slightly thickens.
5. Stir in parsley and green onions.
6. Serve each portion with one cup of cooked **Toro Brand Rice.**

Yields 6 servings.

CRAWFISH DUMPLING ÉTOUFFÉE

I am always trying to create dishes that utilize traditional ingredients in a unique fashion. I think this is one of those dishes.

DUMPLINGS

- *½ pound crawfish tails, coarsely chopped*
- *1½ cups Bisquick Baking Mix*
- *¼ cup milk*
- *2 tablespoons chopped green onions*
- *1 tablespoon minced parsley*
- *1 teaspoon Tony Chachere's Creole Seasoning*
- *1 teaspoon garlic powder*
- *1 teaspoon onion powder*

ÉTOUFFÉE

- *4 tablespoons cooking oil*
- *½ cup chopped onion*
- *¼ cup chopped green bell pepper*
- *¼ cup chopped celery*
- *1 tablespoon minced garlic*
- *3 tablespoons flour*
- *2½ cups chicken broth*
- *1 teaspoon paprika*
- *1 teaspoon Tony Chachere's Creole Seasoning*
- *½ pound crawfish tails*

DIRECTIONS

DUMPLINGS

1. Combine all ingredients for dumplings in a bowl and stir until dough forms.
2. Place in refrigerator for 1 hour.

ÉTOUFFÉE

1. In a saucepot, heat oil. Add onion, green bell pepper, celery and garlic and cook for 3 minutes.
2. Stir in flour and cook for an additional minute, being careful not to brown flour.
3. Whip in chicken broth, paprika and Creole seasoning. Lower fire and simmer for 30 minutes.
4. Stir in crawfish and bring back to simmer. Add dumplings one heaping tablespoon at a time until all of dough is utilized. Cook uncovered for 10 minutes. Cover and cook for an additional 10 minutes.

Yield 4 servings.

CRAWFISH-MUSHROOM CASSEROLE

This is a great dish to prepare for your Sunday dinner. It can be put together the day before and then just popped into the oven the next day.

- 1 *tablespoon olive oil*
- ½ *cup chopped onion*
- ½ *cup chopped red bell pepper*
- ¼ *cup chopped green bell pepper*
- ¼ *cup chopped celery*
- 1 *tablespoon minced garlic*
- 2 *cups sliced fresh mushrooms*
- 1 *cup heavy cream*
- 1 *pound crawfish tails*
- 1 *(8-ounce) can* ***Campbell's Cream of Mushroom Soup***
- ¼ *cup chopped green onions*
- 2 *teaspoons Tony Chachere's Creole Seasoning*
- ½ *teaspoon hot sauce*
- ½ *cup Italian bread crumbs*
- ½ *cup shredded cheddar cheese*
- ½ *cup shredded mozzarella cheese*

DIRECTIONS

1. Preheat over to 350°.
2. In a skillet, heat olive oil. Add onion, bell peppers and celery and cook for 3 minutes.
3. Add garlic and mushrooms; cook for additional minute. Stir in heavy cream and bring to a boil. Reduce heat and simmer for 10 minutes, until cream slightly thickens.
4. In a large bowl stir together crawfish tails, **Campbell's Cream of Mushroom Soup,** green onions, Creole seasoning and hot sauce. Stir in vegetable-cream mixture until completely blended.
5. Pour into casserole dish cover and bake for 30 minutes.
6. Remove casserole from oven and top with bread crumbs and cheese. Return to oven and bake uncovered for additional 10 minutes.

Yields 8 side servings.

FRIED FROG LEGS AND FETTUCCINE ALFREDO

Louisiana frog legs unfortunately can only be acquired if you know someone who goes out and gigs them during the hunting season. The ones farmed in foreign countries do not compare in flavor but will do in a pinch.

FRIED FROG LEGS

4 cups cooking oil
24 small frog legs
1 tablespoon plus 2 teaspoons Tony Chachere's Creole Seasoning, divided
1 teaspoon hot sauce
2 eggs
2 cup milk
3 cups fish fry

FETTUCCINE ALFREDO

4 cups heavy whipping cream
2 tablespoons chopped green onions
1 tablespoon chopped garlic
1 tablespoon chopped parsley
1 tablespoon Tony Chachere's Creole Seasoning
1 teaspoon dried basil leaves
½ cup grated Parmesan cheese
1 pound fettuccine noodles, cooked according to package directions

DIRECTIONS

FROG LEGS

1. Heat cooking oil to 350°.
2. In a bowl, place frog legs and season with 1 tablespoon of Creole seasoning and hot sauce. In a separate bowl beat together eggs, milk and 1 teaspoon of Creole seasoning.
3. In another bowl, place fish fry and season with remaining Creole seasoning.
4. Dip each frog leg into egg batter, then into fish fry, coating frog leg completely. Carefully drop each frog leg into heated vegetable oil and fry until golden brown; keep in a warm oven while making fettuccine alfredo.

FETTUCCINE ALFREDO

1. In large skillet, stir together heavy cream and all ingredients except cheese and cooked pasta. Simmer until sauce starts to thicken slightly, about 5 minutes. Add cheese and simmer until cheese melts and sauce continues to thicken; reduce heat. Add fettuccine noodles and simmer until noodles are warm.
2. Divide fettuccine alfredo onto 4 plates and top with 6 frog legs.

Yields 4 servings.

GRILLED SOFT-SHELL CRABS WITH LUMP CRABMEAT SAUTÉ

One of my favorite foods is soft-shell crab. Crabs shed their hard outer shell every time they go through a growing spurt; the result is a completely edible crab.

GRILLED CRABS

- *2 jumbo soft-shell crabs*
- *1 tablespoon olive oil*
- *1 tablespoon lemon juice*
- *1 teaspoon Tony Chachere's Creole Seasoning*
- *½ teaspoon granulated garlic*
- *½ teaspoon granulated onion*

LUMP CRAB SAUTÉ

- *2 tablespoons unsalted butter*
- *1 tablespoon minced garlic*
- *1 tablespoon lemon juice*
- *¼ cup white wine*
- *½ teaspoon Tony Chachere's Creole Seasoning*
- *¼ teaspoon hot sauce*
- *½ pound lump crabmeat*

DIRECTIONS

GRILLED CRABS

1. Combine all ingredients in a bowl and marinate soft-shell crabs for 30 minutes. Heat grill and cook for about 5 minutes on each side.
2. Set aside and keep warm while making Lump Crab Sauté.

LUMP CRAB SAUTÉ

1. Add all ingredients to a skillet except for lump crabmeat. Simmer over low heat for 2 minutes.
2. Stir in the lump crabmeat and simmer until thoroughly heated. Be careful not to break up lumps of crabmeat.
3. Top each grilled crab with Lump Crab Sauté.

Yields 2 servings.

LOUISIANA BLUE POINT LUMP CRABMEAT ALFREDO

This classic Italian dish gets a Cajun treatment with the addition of spices and lump crabmeat.

- 2 *tablespoons unsalted butter*
- 2 *tablespoons minced garlic*
- 1 *tablespoon dried basil leaves*
- 2 *teaspoons dried oregano leaves*
- 4 *cups heavy whipping cream*
- ¼ *cup grated Parmesan cheese*
- 1 *tablespoon Tony Chachere's Creole Seasoning*
- 1½ *teaspoons hot sauce*
- ½ *teaspoon salt*
- 1 *pound lump crabmeat*
- 1 *pound fettuccine, cooked according to package directions*
- ¼ *cup chopped green onions*
- 2 *tablespoons minced parsley*

DIRECTIONS

1. In a large saucepot, heat butter until melted. Add garlic, basil and oregano; cook for 1 minute.

2. Add cream, Parmesan cheese, Creole seasoning, hot sauce and salt. Simmer until cream starts to reduce and thicken slightly.

3. Gently stir in crabmeat; simmer for 1 minute. Add pasta and cook until heated.

4. Stir in green onions and parsley and simmer for an additional minute.

Yields 4 servings.

LOUISIANA CRAWFISH AU GRATIN

This sauce is a Cajunized version of a traditional béchamel sauce, rich and creamy. This dish can be prepared ahead of time and just popped into the oven 30 minutes prior to serving.

4 tablespoons unsalted butter
½ cup chopped onions
1 tablespoon minced garlic
3 tablespoons flour
1 cup milk
1 cup shredded cheddar cheese, divided
2 teaspoons Tony Chachere's Creole Seasoning
1 teaspoon hot sauce
½ teaspoon dried thyme leaves
¼ teaspoon nutmeg
1 pound crawfish tails
¼ cup chopped green onions
¼ cup Italian bread crumbs

DIRECTIONS

1. Preheat oven to 350°.
2. In saucepot, heat butter over medium heat; add onions and cook for 5 minutes. Add garlic and cook for an additional minute.
3. Stir in flour and cook for a minute. Slowly add milk and whip until milk starts to thicken. Lower heat and simmer for 5 minutes.
4. Add ½ cup cheese, Creole seasoning, hot sauce, thyme and nutmeg. Continue to cook over low heat until cheese is melted.
5. Remove from heat and stir in crawfish tails and green onions. Pour into 8 x 8-inch baking dish and top with remaining cheese and bread crumbs.
6. Bake for 30 minutes.

Yields 4 servings.

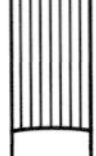

MARINATED GRILLED CATFISH

The best way to determine if the catfish is cooked is if a knife easily pierces the thickest part of the catfish.

4 teaspoons wine vinegar
1 tablespoon Dijon mustard
1 tablespoon minced garlic
1 tablespoon dried basil leaves
1½ teaspoons dried thyme leaves
1 teaspoon Tony Chachere's Creole Seasoning
1 teaspoon hot sauce
¼ teaspoon salt
½ cup olive oil
4 (8-ounce) catfish fillets

DIRECTIONS

1. In the bowl of a food processor or in a mixing bowl, combine all ingredients except olive oil and catfish.
2. Whip together all ingredients until blended; slowly pour in olive oil in a steady stream until incorporated.
3. Add catfish and coat with marinade. Marinate catfish under refrigeration for 2 hours.
4. Fire up outdoor grill using **Royal Oak Charcoal.** Grill catfish for 5 minutes on each side, being careful not to overcook.

Yields 4 servings.

PAN SAUTÉED FLOUNDER WITH OYSTER MEUNIÈRE SAUCE

Meunière Sauce is a classic sauce that can be found in some of the finest restaurants in New Orleans. The sauce is thickened slightly by adding chilled butter just before you are ready to serve; make sure the fire is extremely low during this process.

PAN SAUTÉED FLOUNDER

- 3 *tablespoons olive oil*
- 1 *egg*
- ½ *cup milk*
- ¼ *teaspoon hot sauce*
- 1 *cup flour*
- 1½ *teaspoons Tony Chachere's Creole Seasoning, divided*
- 2 *(8-ounce) flounder fillets*

OYSTER MEUNIÈRE

- 1 *teaspoon minced garlic*
- ¼ *cup white wine*
- 1 *teaspoon Worcestershire sauce*
- 1 *teaspoon lemon juice*
- 12 *oysters*
- 3 *tablespoons chilled unsalted butter*
- 1 *tablespoon chopped green onions*
- 1 *tablespoon minced parsley*

DIRECTIONS

PAN SAUTÉED FLOUNDER

1. Heat olive oil in medium skillet.
2. In a bowl, beat together egg, milk, hot sauce and ½ teaspoon Creole Seasoning. In another bowl toss together flour and remaining Creole Seasoning.
3. Dip flounder fillets in milk-egg batter, then coat in flour mixture.
4. Place flounder in heated oil and cook for 3 minutes until brown. Turn and cook on other side until brown. Remove from pan and keep in a warm place.

OYSTER MEUNIÈRE SAUCE

1. Pour off excess oil, leaving browned flour in bottom of pan. Lower fire and add garlic. Cook for 1 minute, being careful not to burn. Add minced garlic, white wine, Worcestershire and lemon juice, and simmer until wine is reduced by half.
2. Add oysters and cook for 1 minute. Lower heat and stir in chilled butter until melted. Sauce will thicken slightly. Add green onions and parsley. Top each flounder with equal amounts of oysters and sauce.

Yields 2 servings.

ORIENTAL FISH FRY

I was inspired to create this dish by Oriental cuisine, one of my favorites. I am also a big fan of contrasting textures, of which this recipe is a good example. The crunchiness of the vegetables works particularly well with the whole fried catfish.

FISH FRY

- *3 cups vegetable oil*
- *1 cup flour*
- *1 teaspoon salt, divided*
- *1 teaspoon black pepper, divided*
- *1 egg*
- *1 cup milk*
- *1 cup fish fry*
- *2 (8-ounce) whole, bone-in catfish*

VEGETABLE STIR-FRY

- *1 tablespoon olive oil*
- *¼ cup sliced onion*
- *¼ cup sliced celery*
- *¼ cup sliced squash*
- *¼ cup sliced zucchini*
- *1 tablespoon minced fresh ginger*
- *1 cup shredded Napa cabbage*
- *½ cup snow peas*
- *½ cup sliced red bell peppers*
- *½ cup sliced water chestnuts*
- *½ cup bamboo shoots*
- *6 ears baby corn*
- *2 tablespoons soy sauce*
- *½ teaspoon salt*
- *¼ teaspoon white pepper*
- *1 cup chicken broth*
- *1 tablespoon cornstarch*
- *¼ cup sliced green onions*

DIRECTIONS

1. Heat oil to 350° F.
2. In a bowl, place flour and season with ½ teaspoon salt and ½ teaspoon black pepper. In another bowl, beat together egg and milk. In another bowl, place fish and season with remaining salt and pepper.
3. Roll catfish in flour, then dip in milk-egg batter and then roll in fish fry. Carefully place battered fish in oil and cook until golden brown and floating; cook for additional minute for extra crunchiness. Set aside while making stir-fry.

VEGETABLE STIR-FRY

1. Heat olive oil in a large skillet; add onion and cook for 2 minutes. Add celery, squash, zucchini and ginger; continue to cook for another minute.

Flying Gaspergoo

2. Add Napa cabbage, snow peas, red bell peppers, water chestnuts, bamboo shoots, baby corn, soy sauce, salt and pepper.

3. In a cup, stir together chicken broth and cornstarch until dissolved. Stir into vegetables. Cook until sauce starts to thicken, about 2 minutes. Stir in green onions.

4. Divide vegetables between 2 plates and top with catfish.

Yields 2 servings.

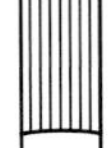

PAT'S CRAWFISH ÉTOUFFÉE

Crawfish consumption really gets into high gear during the Lenten season. Growing up in Crowley, Louisiana, which is surrounded by rice fields converted to crawfish ponds during the off-season, there was no shortage of these tasty crustaceans.

- 1 *stick unsalted butter*
- 3 *cups minced onions*
- 1 *tablespoon minced garlic*
- 4 *teaspoons flour*
- 1½ *teaspoons dried thyme leaves*
- 1 *cup chicken broth*
- 1 *tablespoon Tony Chachere's Creole Seasoning*
- 1 *teaspoon paprika*
- 1 *teaspoon hot sauce*
- 1 *pound crawfish tails*
- ¼ *cup chopped green onions*
- 2 *tablespoons chopped parsley*
- 4 *cups cooked* **Toro Brand Rice**

DIRECTIONS

1. In a saucepot, heat butter over medium heat. Cook onions and garlic for five minutes.
2. Add flour and thyme; cook for 1 minute, being careful not to brown flour.
3. Add chicken broth, Creole seasoning, paprika and hot sauce. Cook for an additional 2 minutes.
4. Stir in crawfish, cover, lower heat and simmer for 10 minutes, stirring occasionally.
5. Stir in green onions and parsley. Serve with cooked **Toro Brand Rice.**

Yields 4 servings.

PECAN-CRUSTED CATFISH, TOPPED WITH SPICY LEMON-BEURRE BLANC

By coating the catfish in the pecans you achieve a crunchy and nutty exterior that is a unique change from the more traditional coating of flour or bread crumbs. The spicy lemon butter sauce is a nice complement to the fish.

- 1 *cup flour*
- 1 *tablespoon Tony Chachere's Creole Seasoning, divided*
- 1 *teaspoon salt, divided*
- 1 *egg*
- ½ *cup milk*
- 2 *(8-ounce) catfish fillets*
- ½ *cup pecan pieces*
- 3 *tablespoons unsalted butter*

SPICY LEMON-BEURRE BLANC

- 4 *lemon slices*
- ½ *cup white wine*
- 3 *tablespoons chilled unsalted butter*
- 1 *teaspoon Tony Chachere's Creole Seasoning*
- ½ *teaspoon hot sauce*
- ¼ *teaspoon salt*

DIRECTIONS

1. Place flour in a bowl and season with half of Creole seasoning and salt. In another bowl beat together egg and milk with remaining Creole seasoning and salt. In another bowl, place pecan pieces.
2. Roll catfish fillets one at a time in seasoned flour, then dip into egg-milk batter. Then dredge one side of fish into pecan pieces and dredge other side back into flour.
3. Heat 3 tablespoons butter in a skillet over medium heat. Place catfish pecan-side-down in skillet and cook 5 minutes, until browned. Turn over and brown on other side. Cook an additional 5 minutes.
4. Remove from skillet and set aside while making Beurre Blanc in same skillet.

SPICY LEMON-BEURRE BLANC

1. Add lemon slices to skillet and cook for 1 minute. Add white wine and cook until reduced by half.
2. Lower fire and stir in butter, Creole seasoning, hot sauce and salt. Cook until butter melts.
3. Top Pecan-Crusted Catfish with equal amounts of butter sauce.

Yields 2 servings.

SEAFOOD AND TASSO PASTA

My son Ethan was 5 years old the first time we cooked this dish together, and well on his way to a lifetime of appreciating good food.

- *1 tablespoon olive oil*
- *1 cup diced tasso*
- *¾ cup chopped onion*
- *½ cup chopped green bell pepper*
- *½ cup chopped red bell pepper*
- *¼ cup chopped celery*
- *2 tablespoons minced garlic*
- *1 tablespoon dried basil leaves*
- *1 tablespoon Tony Chachere's Creole Seasoning*
- *1½ teaspoons dried thyme leaves*
- *¼ teaspoon hot sauce*
- *2 cups heavy whipping cream*
- *½ cup grated Parmesan cheese*
- *½ pound medium shrimp, peeled*
- *½ pound crawfish tails*
- *½ pound rotini pasta, cooked*
- *¼ cup chopped green onions*
- *¼ cup chopped parsley*

DIRECTIONS

1. In a large saucepot, heat olive oil over medium heat. Add tasso and cook until tasso starts to brown slightly.
2. Add onion, bell peppers, celery and garlic; cook for 5 minutes, stirring occasionally. Add basil, Creole seasoning, thyme and hot sauce; cook an additional minute.
3. Add cream and bring to a boil. Reduce heat to a simmer and add Parmesan cheese. Cook for 3 minutes or until sauce starts to thicken slightly.
4. Add shrimp and crawfish tails and simmer for 5 minutes. Add cooked pasta and simmer until pasta is hot.
5. Stir in green onions and parsley and serve.

Yields 6 servings.

SHRIMP, MIRLITON AND CORN MACQUE CHOUX

Mirlitons are also referred to as vegetable pears or chayote squash. You can find them growing on vines in Louisiana backyards. Their delicate flavor generally absorbs the taste of other foods they contact. They are also used in Caribbean, Latin and Southwestern American dishes.

- *3 mirlitons*
- *4 cups water*
- *4 tablespoons unsalted butter*
- *3 ears of corn, shucked and kernels cut off cob*
- *1 cup chopped onion*
- *½ cup diced celery*
- *½ cup diced green bell pepper*
- *1 tablespoon minced garlic*
- *1½ teaspoons dried thyme leaves*
- *1 (10-ounce) can diced Rotel tomatoes, drained*
- *1 tablespoon Tony Chachere's Creole Seasoning*
- *1 pound small shrimp, peeled*
- *2 cups heavy whipping cream*
- *¼ cup grated Parmesan cheese*
- *¼ cup chopped green onions*
- *¼ cup minced parsley*
- *6 cups cooked rice, pasta or polenta*

DIRECTIONS

1. Place mirlitons in a saucepot and cover with water; bring to boil and cook until tender. Cool under running water, peel, seed and cut into ½-inch cubes; set aside for use later. Heat butter in saucepot. Add corn and cook 10 minutes.
2. Add onion, celery, bell pepper, garlic and thyme. Cook for additional 5 minutes.
3. Add Rotel tomatoes and Creole seasoning and cook for 10 minutes, until water from the tomatoes has evaporated.
4. Stir in shrimp, heavy cream, Parmesan cheese and cooked mirliton. Bring to boil. Reduce heat and simmer for 10 minutes, until cream thickens slightly.
5. Stir in green onions and parsley.
6. Serve each portion with one cup of cooked rice, pasta or polenta.

Yields 6 servings.

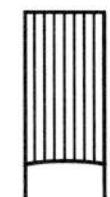

SHRIMP AND EGG STEW

The tradition of topping this dish with boiled eggs came about as a way of stretching the shrimp in the dish.

1 tablespoon olive oil
1 cup chopped onions
½ cup chopped green bell pepper
½ cup chopped red bell pepper
½ cup chopped celery
1 tablespoon minced garlic
1 tablespoon Tony Chachere's Creole Seasoning
1 teaspoon dried thyme leaves
6 cups chicken broth
1 cup ***Savoie's Dry Roux***
1 pound medium shrimp, peeled and deveined
¼ cup chopped green onions
2 tablespoons minced parsley
6 cups cooked Toro Brand Rice
6 boiled eggs, diced

DIRECTIONS

1. Heat olive oil in medium saucepot. Add onion, bell peppers and celery; cook for 5 minutes.
2. Stir in garlic, Creole seasoning and thyme. Continue to cook for 5 minutes.
3. Add chicken broth. Bring to boil and whip in **Savoie's Dry Roux** until dissolved.
4. Simmer for 30 minutes until sauce starts to thicken.
5. Stir in shrimp and simmer for 10 minutes. Stir in green onions and parsley.
6. Serve over cooked Toro Brand Rice; top each serving with diced boiled egg.

Yields 6 servings.

SHRIMP AND TASSO JAMBALAYA

The predecessor to jambalaya was the Spanish paella. The great thing about jambalaya is that it can be made with whatever ingredients are available; in this case, I have used shrimp and smoked tasso.

- *2 tablespoons cooking oil*
- *½ pound chopped tasso*
- *1 cup chopped onion*
- *½ cup chopped green bell pepper*
- *½ cup chopped celery*
- *1 (14-ounce) can diced tomatoes, drained*
- *3 tablespoons dark roux*
- *2 tablespoons minced garlic*
- *2 teaspoons Tony Chachere's Creole Seasoning*
- *1 tablespoon paprika*
- *1 teaspoon salt*
- *1 teaspoon hot sauce*
- *2 teaspoons dried thyme leaves*
- *3½ cups chicken broth*
- *2 cups uncooked* ***Toro Brand Rice***
- *1 pound medium shrimp, peeled and deveined*
- *¼ cup minced parsley*
- *¼ cup chopped green onions*

DIRECTIONS

1. In a large saucepot, heat oil. Add tasso and cook until tasso is browned; add onion, bell pepper and celery. Cook for one minute.
2. Stir in diced tomatoes, roux, garlic, Creole seasoning, salt, paprika and hot sauce. Cook for 10 minutes, until roux is dissolved. Add thyme and continue to cook for 2 minutes.
3. Add chicken broth and bring to boil. Add **Toro Brand Rice** and shrimp. Bring back to boil, lower fire, and cover.
4. Cook for 30 minutes, stirring every 10 minutes. Turn off fire and stir in parsley and green onions. Allow to sit for 5 minutes before serving.

Yields 6 servings.

SHRIMP AND TASSO SAUTÉ

My friend Floyd Poché and his ancestors have been making tasso, the smoked pork delicacy, since 1859 on the banks of the Bayou Teche just outside of Breaux Bridge, Louisiana. Floyd continues the tradition at the family store, shipping smoked meats throughout the world.

Nonstick pan spray
1 *cup chopped tasso*
½ *cup minced onion*
¼ *cup minced green bell pepper*
¼ *cup minced celery*
1 *tablespoon minced garlic*
1 *pound medium shrimp, peeled and deveined*
1½ *cups sliced mushrooms*
½ *cup chicken broth*
½ *cup white wine*
1 *tablespoon Tony Chachere's Creole Seasoning*
1 *teaspoon hot sauce*
¼ *cup chopped green onions*
¼ *cup chopped parsley*
4 *cups cooked* ***Toro Brand Rice***

DIRECTIONS

1. Spray a large skillet with pan spray and place over high heat.
2. Add tasso, onion, bell pepper, celery and garlic. Cook for 5 minutes, browning slightly.
3. Add shrimp, mushrooms, chicken broth, white wine, Creole seasoning and hot sauce, cooking for 5 minutes over high heat.
4. Stir in green onions and parsley; simmer for additional minute.
5. Serve over cooked **Toro Brand Rice.**

Yields 4 servings.

SHRIMP CREOLE

The key to this recipe is not cooking it too long. You want the vegetables to remain crunchy, and you'll also avoid overcooking the shrimp.

2 tablespoons unsalted butter
1 cup chopped onion
½ cup chopped celery
½ cup chopped green bell pepper
½ cup chopped red bell pepper
1 tablespoon minced garlic
1 teaspoon dried thyme leaves
1 tablespoon Tony Chachere's Creole Seasoning
½ teaspoon hot sauce
2 cups chicken broth
1 (14-ounce) can diced tomatoes, drained
1 (8-ounce) can tomato sauce
1 (6-ounce) can tomato paste
1 cup tomato ketchup
1 lemon, sliced
2 bay leaves
2 pounds medium shrimp, peeled and deveined
¼ cup minced green onions
¼ cup minced parsley
8 cups cooked ***Toro Brand Rice***

DIRECTIONS

1. In a large saucepot, heat butter. Add onion, celery, bell peppers and cook for 2 minutes.
2. Stir in garlic, thyme, Creole seasoning and hot sauce. Cook an additional minute. Add chicken broth, diced tomatoes, tomato sauce, tomato paste, ketchup, lemon slices and bay leaves. Bring to simmer and cook for 30 minutes.
3. Add shrimp, simmer for 10 minutes. Be careful not to overcook shrimp.
4. Stir in green onions and parsley. Ladle serving around 1 cup of cooked **Toro Brand Rice.**

Yields 8 servings.

SOUTHWESTERN BARBECUE SHRIMP AND PASTA

These shrimp are also great just eating out of the pan while standing around the stove. Just make sure you have plenty of wet towels handy.

- 1 *pound large shrimp, peeled and deveined*
- 1 *teaspoon Tony Chachere's Creole Seasoning*
- ¼ *teaspoon chili powder*
- ⅛ *teaspoon cumin*
- 2 *tablespoons unsalted butter*
- 1 *teaspoon minced garlic*
- ¼ *cup dried chiles, soaked in water for 15 minutes, drained and thinly sliced*
- ¼ *cup* ***Jack Miller's Barbecue Sauce***
- 3 *tablespoons chicken broth*
- 2 *tablespoons minced cilantro*
- 1 *tablespoon Worcestershire sauce*
- 2 *cups your favorite cooked pasta*

DIRECTIONS

1. In a bowl, toss together shrimp with Creole seasoning, chili powder and cumin. Set aside.
2. In skillet, heat butter. Add garlic and chiles, and cook until browned. Add shrimp and continue to cook over high heat for 2 minutes.
3. Add **Jack Miller's BBQ Sauce,** chicken broth, cilantro and Worcestershire sauce. Simmer for 3 minutes.
4. Serve over your favorite pasta with a crusty loaf of French bread.

Yields 2 servings.

MEATS

MEATS

BRAISED OSSOBUCO WITH TARRAGON BORDELAISE

Veal shank is a very affordable cut of meat that is under-utilized, in my opinion. The key to this dish is a long, slow-cooking process; it will be so tender it will melt in your mouth.

- *1 pound crosscut veal shank*
- *2 teaspoons Tony Chachere's Creole Seasoning, divided*
- *1 cup plus 1 tablespoon flour, divided*
- *2 tablespoons olive oil*
- *1 cup chopped onions*
- *2 tablespoons tomato paste*
- *1 tablespoon minced garlic*
- *2½ cups beef broth*
- *¼ cup red wine*
- *2 teaspoons dried tarragon*
- *2 tablespoons chopped green onion*
- *1 tablespoon chopped parsley*

DIRECTIONS

1. Preheat oven to 350°.
2. Season veal shank with 1 teaspoon of Creole seasoning. In a bowl, toss together remaining Creole seasoning and cup of flour.
3. In an ovenproof saucepot, heat olive oil. Dredge veal shank through seasoned flour and place in pot. Cook until browned on both sides. Remove from the saucepot and set aside.
4. Add onion, tomato paste, garlic and remaining tablespoon of flour. Cook for 2 minutes. Add beef broth, red wine and tarragon. Whip until sauce is formed.
5. Place veal shank back in the pot and cover. Place in oven and bake for 1 hour. Stir twice during cooking process.
5. Remove from the oven and serve with your favorite pasta or rice dish.

Yields 2 servings.

BREAST OF CHICKEN DIABLO

Don't be startled by the amount of black pepper in this dish. It's a nice accompaniment to the chicken, but you can always use less.

- *4 (6-ounce) boneless chicken breasts*
- *4 tablespoons cracked black pepper, divided*
- *3 tablespoons unsalted butter*
- *¼ cup chopped onions*
- *¼ cup chopped green onions*
- *1 tablespoon minced garlic*
- *2 tablespoons flour*
- *1¼ cups chicken broth*
- *¼ cup white wine*
- *1 teaspoon salt*

DIRECTIONS

1. Season each chicken breast with 1 tablespoon cracked black pepper. In a skillet, heat butter over medium heat; allow butter to brown slightly. Add chicken breasts and cook for 2 minutes on each side until chicken breasts are browned. Remove from pan and keep warm.

2. Add onion and green onions and cook for 2 minutes. Add garlic and cook for additional minute.

3. Add flour and cook until flour starts to brown slightly, approximately 2 minutes. Stir in chicken broth and white wine. Simmer until sauce starts to thicken slightly.

4. Return chicken breasts to sauce. Stir in salt and simmer for 10 minutes until chicken is cooked.

5. Place chicken breasts on platter and top with sauce.

Yields 4 servings.

CHICKEN AND VEGETABLE STIR FRY

This recipe is a dieter's dream! The fresh ginger and chicken broth add a lot of low-fat flavor. Be especially careful not to overcook the vegetables.

2 teaspoons rice vinegar
2 teaspoons lime juice
1 teaspoon teriyaki sauce
1 teaspoon soy sauce
1 teaspoon minced ginger, divided
½ teaspoon salt
1¼ pounds boneless skinless chicken breast strips
Nonstick pan spray
1¼ cups chicken broth, divided
2 teaspoons cornstarch
1 cup broccoli florets
1 cup cauliflower florets
½ red bell pepper, julienned
½ green bell pepper, julienned
1 teaspoon minced garlic
2 cups shredded Napa cabbage
½ cup sliced water chestnuts
½ teaspoon salt
½ teaspoon white pepper
¼ cup sliced green onions
4 cups cooked brown rice

DIRECTIONS

1. In a bowl, beat together rice vinegar, lime juice, teriyaki sauce, soy sauce, ½ teaspoon ginger and salt. Add chicken breast strips and marinate in the refrigerator for up to 2 hours.
2. Heat a large skillet and spray with nonstick pan spray. Add chicken strips and cook, browning slightly for several minutes. Remove from skillet and place in clean bowl; set aside. Whisk ¼ cup of chicken broth into sauté pan, scraping bottom of pan. Reduce by half and then pour into bowl with chicken.
3. Spray skillet with more nonstick pan spray and return to medium heat. Add broccoli and cauliflower and cook for one minute stirring constantly.
4. Add bell peppers, remaining ginger and garlic and continue to cook for 1 minute. Add cabbage, water chestnuts, salt and white pepper.
5. In a small bowl combine remaining cup of chicken broth and cornstarch; stir until cornstarch is dissolved. Stir into skillet, cover and cook for 1 minute. Sauce will thicken. Lower fire, add chicken strips and simmer until chicken strips are heated thoroughly.
6. Stir in sliced green onions. Serve each portion with a cup of brown rice.

Yields 4 servings.

CHICKEN AND DUMPLINGS

Chicken and dumplings is one of greatest recipes the Southern United States has produced. The trick to good dumplings is to make sure your dough is stiff enough to stay together, and to apply a gentle heat during the cooking process.

1 whole raw chicken, cup up 8 ways
¼ gallon of water
2 tablespoons Tony Chachere's Creole Seasoning, divided
¼ cup oil
1½ cups chopped onion
¾ cup chopped celery
¾ cup chopped green bell pepper
1 tablespoon minced garlic
¾ cup flour
8 cups reserved chicken stock
2 cups Bisquick Baking Mix
⅔ cup milk
3 tablespoons chopped green onions
2 tablespoons minced parsley

DIRECTIONS

1. In a large stock pot combine chicken, water and 1 tablespoon Creole seasoning. Bring to boil, lower fire and simmer for 40 minutes. Remove chicken from pot and cool. Debone chicken, discarding skin. Dice meat into medium chunks. Set aside. Reserve 8 cups of chicken stock, adding water if needed to make 8 cups.
2. In a large sauce pot, heat oil over medium fire. Add onion, celery, bell pepper and garlic. Cook for 5 minutes.
3. Add flour and stir until roux has formed. Turn fire to medium high and cook for 2 minutes. Add reserved chicken stock and remaining Creole seasoning and whip until smooth. Simmer for 5 minutes.
4. In a bowl, beat together biscuit mix and milk until dough is formed and set aside.
5. Stir diced chicken into pot and bring back to boil. Reduce heat to simmer. Using a large soup spoon, drop golf ball-sized dumplings into pot one at a time.
6. Simmer dumplings for 10 minutes then flip them over. Cover pot and cook for 5 more minutes.
7. Stir in green onions and parsley.

Yields 6 servings.

CHICKEN AND TASSO JAMBALAYA

The "jam" in jambalaya comes from the French word "jambon" which means ham.

- 2 *tablespoons cooking oil*
- 1 *pound raw boneless chicken thigh meat, cubed*
- ¾ *cup chopped onion*
- ¼ *cup chopped celery*
- ¼ *cup chopped bell pepper*
- 1 *tablespoon garlic*
- ½ *pound chopped tasso*
- 1 *(8-ounce) can Rotel Tomatoes*
- 1 *teaspoon Tony Chachere's Creole Seasoning*
- 1 *teaspoon dried thyme leaves*
- 1 *teaspoon salt*
- 1 *teaspoon paprika*
- 3½ *cups chicken broth*
- 2 *cups uncooked* ***Toro Brand Rice***
- ¼ *cup minced parsley*
- ¼ *cup chopped green onions*

DIRECTIONS

1. In large saucepot, heat oil and cook chicken thigh meat until browned. Add onions, celery, bell pepper and garlic; cook for 5 minutes, stirring occasionally.
2. Add tasso, Rotel Tomatoes, Creole seasoning, thyme, salt and paprika; simmer for 10 minutes.
3. Add chicken broth; bring mixture to a boil.
4. Stir in the **Toro Brand Rice;** bring back to boil, then lower the heat and cover the pot.
5. Cook jambalaya for 30 minutes, stirring every 10 minutes.
6. Add green onions and parsley.

Yields 6 servings.

DR. MILLER'S OVEN-ROASTED BARBECUE CHICKEN

Adding the soft drink Dr. Pepper is a trick my old high school buddy, Ned Fowler, shared with me. It's one of those ingredients that people may not be able to identify, but they will know something tastes just a little different about the dish.

TARRAGON BUTTER

- 4 *tablespoons unsalted butter*
- ¾ *teaspoon dried tarragon*
- ½ *teaspoon lemon juice*
- ¼ *teaspoon hot sauce*
- ¼ *teaspoon Tony Chachere's Creole Seasoning*
- ¼ *teaspoon chili powder*
- 1 *(3½ to 4-pound) whole chicken*

BARBECUE MARINADE

- ½ *cup* ***Jack Miller's Barbecue Sauce***
- ¼ *cup Dr. Pepper soft drink*
- 3 *tablespoons honey*
- ½ *teaspoon salt*
- 1 *cup chicken broth*

DIRECTIONS

TARRAGON BUTTER

1. Preheat oven to 375°. Allow butter to soften slightly at room temperature. It should still have a slight chill to it.
2. Place butter, tarragon, lemon juice, hot sauce, Creole seasoning and chili powder in small bowl. Mash with a fork until blended.
3. Divide butter into four equal amounts. Loosen breast skin of chicken away from meat and stuff with a quarter of tarragon butter on one side of breast; repeat method for other half of chicken using another quarter. Loosen skin in leg and thigh area and stuff another fourth of butter between skin and meat; repeat method on other leg and thigh area with remaining amount of butter. Set aside.

BARBECUE MARINADE

1. In a bowl large enough to hold the chicken, combine all ingredients for marinade, except chicken broth. Stir until blended. Place chicken in bowl and roll in marinade. Marinate

from 4 hours to overnight in the refrigerator. Rotate occasionally.

2. Place chicken in a roasting pan with chicken broth. Set aside leftover marinade in bowl. Place chicken in oven.

3. Bake for 1 hour, basting every 20 minutes with marinade.

4. Lower heat to 325° and bake for 40 minutes. Add additional chicken broth if pan begins to dry out.

5. Remove chicken from oven and let stand at room temperature for 20 minutes before carving.

Yields 4 servings.

GRILLED GARLIC PEPPER COWBOY STEAKS

There is no better cut of meat then a bone-in rib-eye, which is also referred to as a cowboy steak. The meat closest to the bone always has the best flavor. Your local butcher will be more than happy to cut you some of these steaks.

2 tablespoons olive oil
1 teaspoon of dried tarragon
1½ teaspoons Worcestershire sauce
1½ teaspoons minced garlic
1 teaspoon coarse ground black pepper
¾ teaspoon chili powder
¼ teaspoon salt
2 (16-ounce) bone-in rib-eyes

DIRECTIONS

1. In a bowl, combine all ingredients except rib-eye steaks until paste forms.
2. Rub paste into steak and marinate in refrigerator for 1 hour.
3. Fire up Bar-B-Que pit using **Royal Oak Charcoal** and grill steaks to desired doneness over hot fire.

Yields 2 servings.

OLD-FASHIONED CHICKEN POT PIE

Utilizing the Bisquick Biscuit Mix as an ingredient adds to the preparation ease of this recipe. And it tastes good, too!

- *½ raw bone-in chicken, cut up*
- *4 cups water*
- *2½ teaspoons Tony Chachere's Creole Seasoning, divided*
- *1 tablespoon olive oil*
- *1 cup chopped onion*
- *1 cup chopped celery*
- *1 cup chopped bell pepper*
- *1 tablespoon minced garlic*
- *1 teaspoon celery seed*
- *½ teaspoon salt*
- *¼ teaspoon white pepper*
- *1 cup diced cooked carrots*
- *1 cup diced cooked potatoes*
- *1 cup sliced mushrooms*
- *4 cups reserved chicken stock (add water if necessary to get 4 cups)*
- *5 tablespoons cornstarch*
- *5 tablespoons water*
- *¼ cup chopped green onions*
- *¼ cup minced parsley*
- *2 cups Bisquick Baking Mix*
- *1 cup milk*
- *1 egg*

DIRECTIONS

1. Preheat oven to 400°. In a large stock pot place chicken, water and 1½ teaspoons Creole seasoning; bring to a boil. Reduce heat and simmer until chicken is cooked, about 40 minutes. Strain stock and reserve for recipe. Cool, debone and dice chicken, discarding skin. Set aside.

2. Heat oil in a saucepot over medium heat. Add onions, celery and bell pepper and cook for 5 minutes. Add garlic, celery seed, remaining Creole Seasoning, salt and white pepper; cook for 2 minutes.

3. Add carrots, potatoes and mushrooms; cook for 3 minutes. Add reserved chicken stock and bring to a boil. Simmer for 5 minutes.

4. In a cup, stir together cornstarch and water until dissolved. Stir into mixture and cook until it has thickened. Stir in green onions and parsley. Set aside to cool.

5. In a bowl, beat together biscuit mix, milk and egg until dough forms. Roll out dough into an 8 x 10 x 1-inch square.

6. Place chicken pie filling in an 8 x 10-inch baking dish. Place dough over baking dish, pressing down ends of dough to seal top. Cut five one-inch slits in dough to allow steam to escape during baking.

7. Bake for 30 minutes. Remove from oven and cool slightly before serving.

Yields 8 servings.

PANEED BREAST OF CHICKEN

This paneed chicken breast is an excellent accompaniment to the Pasta Macque Choux recipe on page 149.

4 (4-ounce) raw boneless, skinless chicken breasts, pounded thin
3 teaspoons Tony Chachere's Creole Seasoning, divided
½ cup flour
½ cup Italian bread crumbs
1 egg
½ cup milk
¼ cup olive oil
4 lemon slices

DIRECTIONS

1. Season chicken breasts with 2 teaspoons of Creole seasoning.
2. Place flour and bread crumbs in two separate bowls. In another bowl beat together egg, milk and remaining teaspoon of Creole seasoning.
3. Dredge chicken breasts in flour, then dip in egg batter. Roll in bread crumbs, pressing bread crumbs firmly into chicken.
4. In large skillet, heat olive oil. Place chicken breasts in pan and cook on both sides until completely browned and cooked.
5. Place on platter and garnish with lemon slices.

Yields 4 servings.

PINEAPPLE-GLAZED ROASTED HAM

This combination of ingredients will make your Sunday dinner ham a treat for the entire family.

- *1 (10-pound)* ***Bryan Old Fashioned Bone-in Ham***
- *28 whole cloves*
- *3 tablespoons unsalted butter*
- *1 (15-ounce) can crushed pineapple, drained*
- *1¼ cups apple juice, divided*
- *3 tablespoons cane syrup*
- *2 tablespoons pineapple preserves*
- *2 tablespoons Dijon mustard*
- *1 teaspoon Tony Chachere's Creole Seasoning*
- *1 teaspoon minced garlic*
- *1 teaspoon hot sauce*
- *3 cups water*
- *6 pineapple rings*
- *6 red maraschino cherries*
- *6 green maraschino cherries*

DIRECTIONS

1. Preheat oven to 300°. Place **Bryan Old Fashioned Ham** in large roasting pan. Trim smoked ham, leaving ¼ inch of fat covering top. Crisscross cut ham about ¼ inch deep. Stud ham with 24 whole cloves.

2. In small saucepot, heat butter. Add crushed pineapple and remaining cloves. Cook over high heat for 2 minutes until water is evaporated.

3. Stir in ¼ cup apple juice, cane syrup, pineapple preserves, mustard, Creole seasoning, garlic and hot sauce. Continue to cook on high heat until glaze thickens. Spread glaze over top of ham, allowing it to coat sides.

4. Add remaining cup of apple juice and 1 cup of water to bottom of roasting pan. Cover ham, place in oven, cook for 2½ hours, basting every hour. Add 1 cup of water at each basting.

5. Remove from oven and garnish with pineapple rings. Return to oven and cook an additional hour uncovered. Remove from oven and place on serving platter and garnish with cherries.

6. Strain gravy at bottom of pan through a fine mesh strainer and serve with ham.

Yields an Easter Ham for one large family.

SAUSAGE-STUFFED LOIN OF PORK

Stuffing the pork loin with the sausage is not as difficult as you may think. Make sure you have a long thin bladed knife for this process.

- 1 *(2-pound)* ***Bryan Butcher Fresh Boneless Pork Loin***
- 1 *link* ***Bryan Smoked Sausage***
- 1 *tablespoon minced garlic*
- 1 *teaspoon dried rosemary*
- 1 *teaspoon Tony Chachere's Creole Seasoning*
- 1 *teaspoon hot sauce*
- ½ *teaspoon granulated garlic*
- ½ *teaspoon granulated onion*
- ½ *medium onion, minced*
- 2 *stalks celery, cut into chunks*
- ½ *green bell pepper, cut into chunks*
- 2 *cups beef broth*

DIRECTIONS

1. Preheat oven to 350°. Using a long slicing knife, cut into both ends of the **Bryan Butcher Fresh Boneless Pork Loin,** running the knife from the ends to the center. This will create a cavity through the length of the loin.
2. Insert link of **Bryan Smoked Sausage** into pork loin.
3. Season pork loin with garlic, rosemary, Creole seasoning, hot sauce, granulated garlic and granulated onion.
4. Place pork loin in roasting pan and top with onion, celery and bell pepper. Place in oven and roast for 50 minutes or until internal temperature reaches 165°. Remove loin from pan and place on cutting board; allow loin to sit for 10 minutes before slicing into 1-inch thick medallions. Set medallions on serving platter.
5. Place roasting pan over low fire on stove and stir in 2 cups beef broth, scraping vegetables that have stuck to the bottom. Allow to simmer for 5 minutes; strain into serving bowl, removing solids, reserving liquid gravy. Top each serving of pork loin with pan gravy.

Yields 6 servings.

SEARED FAJITA STEAK WITH ROSEMARY-MUSHROOM RED WINE SAUCE

This recipe is a great change from the traditional way of preparing a fajita steak.

2 (10-ounce) beef fajita steaks
1 teaspoon dried rosemary
¾ teaspoon salt, divided
½ teaspoon black pepper
1 tablespoon vegetable oil
2 tablespoons unsalted butter
½ cup chopped onion
1 tablespoon flour
1 teaspoon minced garlic
¾ cup beef broth
¼ cup red wine
2 cups sliced mushrooms
½ teaspoon hot sauce
¼ teaspoon Tony Chachere's Creole Seasoning
1 tablespoon minced green onions
1 tablespoon chopped parsley

DIRECTIONS

1. Season beef fajita steaks with rosemary, ½ teaspoon salt and black pepper.
2. In a large skillet, heat oil. Sear steaks over medium heat until medium doneness. Remove from pan and set aside.
3. Melt butter in skillet; add onions and cook 5 minutes, stirring occasionally. Add flour and garlic and cook 1 minute.
4. Whisk in beef broth and red wine until sauce starts to thicken. Lower fire, add mushrooms, hot sauce, Creole seasoning and remaining ¼ teaspoon salt; continue to cook for 5 minutes.
5. Stir in green onions and parsley. Top each seared fajita steak with sauce.

Yields 2 servings.

SMOKED CORNISH GAME HEN IN A ROSEMARY PORTOBELLO MUSHROOM FRICASSEE

This recipe is just as good if you are unable to smoke the Cornish hens. There may be a smoke house in your neighborhood that would be willing to smoke them for you.

- 2 *(20-ounce) Cornish game hens*
- 2 *tablespoons unsalted butter*
- ¼ *cup minced French shallots*
- 2 *tablespoons minced garlic*
- 1 *teaspoon dried rosemary*
- ¼ *cup red wine*
- ½ *teaspoon Tony Chachere's Creole Seasoning*
- 3½ *cups smoked hen stock*
- ¼ *cup* ***Savoie's Dry Roux***
- 1 *cup sliced portobello mushrooms*
- 6 *cups cooked pasta*

DIRECTIONS

1. Smoke Cornish hens in a conventional smoker the day before. Debone and save bones for stock; discard skin, cut meat into large chunks. Set aside.
2. In a large pot, cover bones with 2 quarts of water and bring to a boil. Lower fire and simmer for 1½ hours. Strain out solids and reserve 3½ cups of stock.
3. In large skillet, heat butter. Add shallots and garlic and cook for 5 minutes.
4. Add rosemary and continue to cook for 2 minutes.
5. Stir in red wine, scraping bottom of pan until wine is reduced by half.
6. Add Creole seasoning and smoked hen stock. Bring to a boil, then whisk in **Savoie's Dry Roux.**
7. Return to a boil, lower fire and simmer for 15 minutes. Add portobello mushrooms and simmer for an additional 15 minutes. Add diced Cornish hen and simmer for 5 minutes.
8. Ladle each serving over 1 cup of cooked pasta.

Yields 6 servings.

SMOKED SAUSAGE CREOLE

Rotel Tomatoes are stewed with chile peppers. If you are unable to locate them in your area, use regular diced tomatoes and just spice up the dish with some extra hot sauce.

- 1 *pound* ***Bryan Smoked Sausage,*** *sliced*
- 1 *cup diced onions*
- ½ *cup diced bell pepper*
- ½ *cup chopped celery*
- 1 *tablespoon minced garlic*
- 1 *cup beef broth*
- 1 *(10-ounce) can diced Rotel tomatoes*
- 1 *(6-ounce) can tomato sauce*
- 1 *tablespoon tomato paste*
- 1 *teaspoon Tony Chachere's Creole Seasoning*
- ½ *teaspoon black pepper*
- ½ *teaspoon dried thyme leaves*
- ¼ *teaspoon salt*
- ½ *cup chopped green onions*
- ½ *cup minced parsley*
- 4 *cups cooked Toro Brand Rice*

DIRECTIONS

1. In a saucepot, brown **Bryan Smoked Sausage.** Add onions, bell pepper and celery, and cook for 5 minutes.
2. Add garlic and cook additional 2 minutes.
3. Stir in beef broth, Rotel Tomatoes, tomato sauce, tomato paste, Creole seasoning, black pepper, thyme and salt. Bring to boil, lower fire, then cook for 30 minutes, stirring occasionally.
4. Uncover and simmer for 15 minutes, until sauce thickens slightly.
5. Stir in green onions and parsley. Ladle each serving over 1 cup of cooked rice.

Yields 4 servings.

SMOKED SAUSAGE AND POTATO FRICASSEE

The potatoes in this dish just soak up the flavor of the sausage and spices.

- 1 *pound* ***Bryan Smoked Sausage,*** *sliced*
- 1 *cup chopped onion*
- ½ *cup chopped green bell pepper*
- ½ *cup chopped celery*
- 1 *tablespoon minced garlic*
- ¼ *cup dark roux*
- 1 *tablespoon Tony Chachere's Creole Seasoning*
- 2 *tablespoons Worcestershire sauce*
- ¼ *teaspoon white pepper*
- 4 *cups beef broth, divided*
- 3 *pounds red potatoes, cubed*
- ¼ *cup chopped green onion*
- 2 *tablespoons chopped parsley*

DIRECTIONS

1. In a large saucepot, brown **Bryan Smoked Sausage.**
2. Add onion, bell pepper, celery and garlic; cook for 2 minutes.
3. Add dark roux, Creole seasoning, Worcestershire sauce and white pepper; cook until roux begins to soften. Add 1 cup beef broth and stir until roux begins to dissolve.
4. Cover pot and simmer for 15 minutes, stirring occasionally.
5. Uncover pot and cook for an additional 5 minutes, stirring occasionally.
6. Add potatoes and remaining beef broth, bring to boil, cover and cook for 20 minutes over medium heat. Remove cover and cook for additional 5 minutes.
7. Add green onions and parsley.

Yields 6 servings.

SMOKED SAUSAGE AND PENNE PASTA AU GRATIN

This is a casserole dish that can be prepared ahead of time in order for you to be able to spend more time with friends and family.

- 1 *pound* ***Bryan Smoked Sausage,*** *sliced*
- 1 *pound penne pasta, cooked according to package directions*
- 3 *cups heavy whipping cream*
- 1 *tablespoon minced garlic*
- 1 *teaspoon dried basil leaves*
- 1 *teaspoon dried oregano leaves*
- ½ *cup diced Rotel tomatoes, drained*
- ¼ *cup minced green onions*
- ¼ *cup minced parsley*
- 2 *teaspoons Tony Chachere's Creole Seasoning*
- 1 *teaspoon hot sauce*
- ½ *cup grated mozzarella cheese, divided*
- ½ *cup grated cheddar cheese, divided*

DIRECTIONS

1. Preheat oven to 350°.
2. Brown sliced **Bryan Smoked Sausage** slightly in skillet.
3. In a bowl, stir together sausage and all remaining ingredients except ¼ cup each of mozzarella and cheddar.
4. Pour ingredients in a casserole dish and top with remaining cheese. Bake for 25 minutes.
5. May be served as a side dish or an entrée.

Yields 12 side dishes or 6 entrée portions.

SMOTHERED ROUND STEAK

My Maw-Maw and I used to walk a short distance from her house in South Crowley, Louisiana, to Robicheaux's Meat Market, where she would get her fresh meat. I was always fascinated by that place . . . all the butchers dressed up in their long white coats handling large sides of beef. As Crowley is also the *Rice Capital of the World,* this dish would not be complete without some rice for the gravy.

- 1 *tablespoon cooking oil*
- 1 *(2-pound) boneless round steak*
- 3 *teaspoons Tony Chachere's Creole Seasoning, divided*
- 2 *teaspoons Worcestershire sauce, divided*
- 2 *cups diced onions, divided*
- ½ *cup diced celery*
- ½ *cup diced green bell pepper*
- 1 *tablespoon minced garlic*
- 3 *cups beef broth*
- 1 *teaspoon hot sauce*
- 2 *tablespoons water*
- 1 *tablespoon cornstarch*
- 6 *cups cooked* ***Toro Brand Rice***

DIRECTIONS

1. In a large pot, heat oil over medium heat. Season each side of round steak with 1 teaspoon of Creole seasoning and 1 teaspoon of Worcestershire sauce.
2. Place steak in saucepot and cook on both sides until completely brown, approximately 20 minutes.
3. Add 1 cup of chopped onion and continue to brown for 10 minutes, turning steak occasionally.
4. Add remaining onions, celery, bell pepper and garlic, and brown for additional 5 minutes.
5. Add beef broth, remaining Creole seasoning, and hot sauce. Cover and simmer for 1 hour.
6. In a cup, stir together water and cornstarch until a smooth paste has formed. Stir into saucepot and cook uncovered until gravy thickens. Place lid on pot, lower fire and cook an additional 15 minutes.
7. Serve each portion with 1 cup of cooked **Toro Brand Rice.**

Yields 6 servings.

69

SPICY MARINARA SAUCE WITH ITALIAN MEATBALLS

I never could get used to having meat sauce instead of meatballs to go with my spaghetti — my mother always made meatballs.

SPICY MARINARA SAUCE

¼ cup olive oil
2 cups chopped onions
1 tablespoon minced garlic
1 tablespoon dried basil leaves
1 teaspoon dried oregano leaves
¼ cup minced flat-leaf Italian parsley
1 bay leaf
2 cups beef broth
1 (10-ounce) can crushed tomatoes
1 (8-ounce) can tomato sauce
1 (6-ounce) can tomato paste
1 teaspoon sugar
¼ teaspoon salt
¼ teaspoon black pepper
16 Italian meatballs, cooked
8 cups cooked pasta

ITALIAN MEATBALLS

1 pound ground beef
1 pound ground turkey
1 cup Italian bread crumbs
½ cup minced onion
½ cup grated Parmesan cheese
2 eggs
2 tablespoons minced flat-leaf Italian parsley
1 teaspoon each of dried basil and oregano
1 teaspoon minced garlic
1 teaspoon Tony Chachere's Creole Seasoning

DIRECTIONS

SPICY MARINARA SAUCE

1. In a sauce pot, heat olive oil. Stir in onions, garlic, basil, oregano, parsley and bay leaf. Cook for 10 minutes, stirring occasionally.
2. Add remaining ingredients except for meatballs and pasta, and simmer for 45 minutes.
3. Add cooked meatballs and simmer for 15 minutes.
4. Serve 2 meatballs over a cup of cooked pasta laced with sauce.

ITALIAN MEATBALLS

1. Preheat oven to 350°.
2. In a bowl, blend together all ingredients.
3. Shape each meatball from ¼ cup of mixture. Place meatballs on a greased baking sheet.
4. Bake for 10 minutes. Remove from oven and loosen with a spatula to prevent sticking. Place back in oven and bake for 20 more minutes.
5. Add to marinara sauce.

Yields 8 servings.

SWEET PEPPER HAM STEAKS WITH FIG GLAZE

You should be able to find Cajun Power and Tiger Sauce on your grocery store shelf. They add a sweet and spicy flavor to the Bryan Ham that is hard to beat.

½ cup Cajun Power Garlic Sauce
½ cup cane syrup
⅓ cup Tiger Sauce
2 tablespoons minced garlic
1 teaspoon onion powder
1 teaspoon hot sauce
¼ teaspoon cayenne
8 (½-inch) slices of ***Bryan Centerpiece Ham***, *6 ounces each*

FIG GLAZE

1 (11-ounce) jar fig preserves
¼ cup chicken broth
2 tablespoons butter
½ teaspoon Tony Chachere's Creole Seasoning
½ teaspoon hot sauce

DIRECTIONS

1. In a bowl, blend together all ingredients except **Bryan Centerpiece Ham** steaks; set aside.
2. Heat large skillet over medium heat.
3. Dip each ham steak in seasoning mixture, coating both sides of steak.
4. Cook ham steaks several minutes on each side until browned.
5. Serve with fig glaze.

FIG GLAZE

1. In small saucepot, stir together all ingredients for fig glaze; bring to boil, lower fire, and simmer until glaze thickens, about 5 to 10 minutes.
2. Serve with ham steaks.

Yields 8 servings.

TEX-MEX BEEF AND BEAN CHILI WITH CREAMY RED PEPPER POLENTA

Polenta is one of my favorite dishes and goes well with this version of chili.

BEEF AND BEAN CHILI

- ½ *pound dried red kidney beans*
- 4½ *cups beef broth, divided*
- 2 *tablespoons olive oil*
- 2 *pounds stew meat, diced small*
- 1½ *teaspoons salt, divided*
- ½ *teaspoon black pepper*
- 1 *cup minced onion*
- 2 *teaspoons minced garlic*
- 1 *(14-ounce) can diced tomatoes, with juice*
- 2 *tablespoons chili powder*
- 1½ *teaspoons ground cumin*
- 1 *(6-ounce) can tomato sauce*
- 1 *medium jalapeño pepper, seeded and minced*
- ½ *pound sliced fresh mushrooms*
- ½ *cup water*
- 1 *tablespoon masa harina or flour*

POLENTA

- 2 *cups water*
- 1 *tablespoon unsalted butter*
- 1 *tablespoon minced garlic*
- 1 *teaspoon salt*
- ¾ *teaspoon crushed red pepper*
- ½ *cup yellow corn meal*

DIRECTIONS

BEEF AND BEAN CHILI

1. Rinse the beans in water and drain, removing any pieces of dirt or rocks. Place beans in large bowl and add 1 cup of beef stock. Soak beans for 1 hour.
2. In a large saucepot, heat oil. Season meat with 1 teaspoon salt and ½ teaspoon black pepper. Add to pot and cook for 15 minutes until brown. Add onion and garlic. Cook for additional 5 minutes.
3. Add diced tomatoes, chili powder and cumin. Cook for 5 minutes. Add red kidney beans, tomato sauce, jalapeño pepper, remaining beef broth and ½ teaspoon salt. Bring to boil, lower heat to simmer, cover and cook for 45 minutes.
4. Add sliced mushrooms and continue to cook, covered, for 15 minutes. Stir occasionally.
5. In a small bowl, stir together water and masa harina (or flour) until masa is dissolved. Stir into chili. Simmer for 2 minutes.
6. Serve with Creamy Red Pepper Polenta on side.

POLENTA

1. In a small saucepot, combine all ingredients except corn meal and bring to boil.
2. Lower fire and slowly pour in corn meal, whipping constantly.
3. Cook until mixture leaves the sides of the saucepot.
4. Serve with Tex-Mex Beef and Bean Chili.

Yields 10 servings.

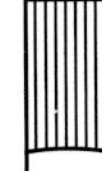

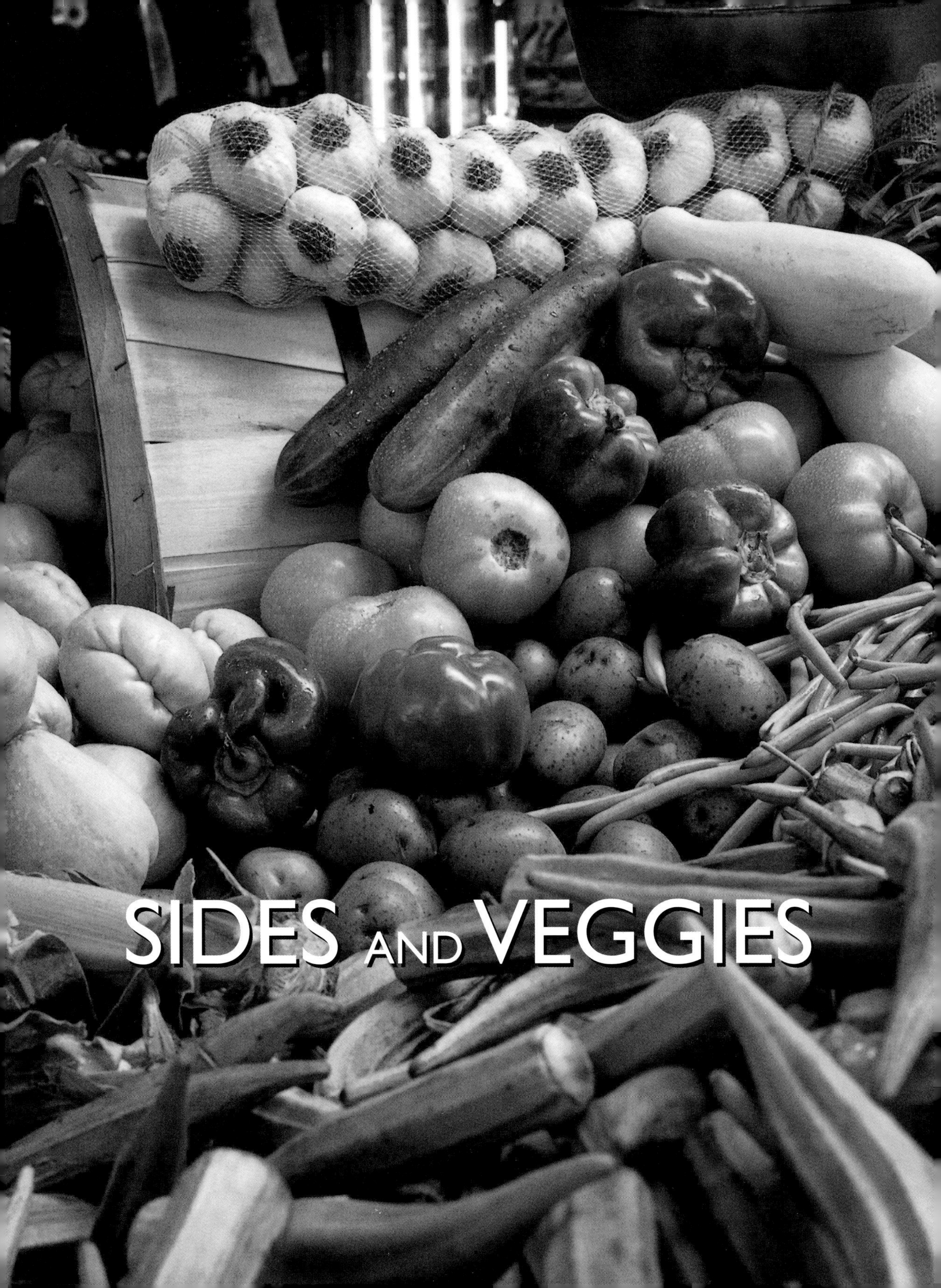

SIDES AND VEGGIES

SIDES AND VEGGIES

Country Fried Okra
Creole Red Beans and Rice
Ham and Asparagus Quiche
Italian Risotto
Louisiana Dirty Rice Dressing
Orzo Pasta Jambalaya
Pan-Fried Eggplant
Pasta Primavera
Pecan Rice
Pork and Eggplant Rice Dressing
Praline Sweet Potato Casserole
Seafood Broccoli Casserole
Seasoned Rice Pilaf
Smoked Andouille Rice Cakes
Smoked Sausage, Shrimp and Eggplant
Smoked White Beans
Tasso and Green Bean Casserole

COUNTRY FRIED OKRA

No summer garden in South Louisiana is planted without including this staple of Cajun-Creole cooking. Cooking the okra slightly in salted water before frying helps to tenderize it.

- 2 *cups water*
- 1 *teaspoon salt*
- 1 *pound whole okra*
- 2 *cups ice water*
- 4 *cups cooking oil*
- 2 *eggs*
- ½ *cup milk*
- ½ *cup buttermilk*
- 1 *teaspoon hot sauce*
- 3 *cups flour*
- 2 *teaspoons Tony Chachere's Creole Seasoning, divided*

DIRECTIONS

1. Combine water and salt in a pot and bring to a boil. Add okra and cook for 2 minutes. Drain and chill in ice water to stop cooking process. Set aside.
2. Heat cooking oil to 350°.
3. In a bowl, beat together eggs, milk, buttermilk and hot sauce.
4. In another bowl, blend together flour and 1 teaspoon of Creole seasoning.
5. Season okra with remaining teaspoon of Creole seasoning.
6. Batter okra by first coating with flour, then the egg batter. Dip back into the flour and completely coat. Shake off any excess flour before frying.
7. Carefully drop okra into heated frying oil and cook until golden brown.

Yields 6 to 8 side servings.

CREOLE RED BEANS AND RICE

I love you once, I love you twice, I love you more than beans and rice!

- 1 *pound dried red beans*
- 1¼ *gallons of beef broth*
- 1 *pound* **Bryan Smoked Sausage,** *sliced*
- 1 *(8-ounce) can diced Rotel tomatoes*
- 2 *cups chopped onions*
- ¾ *cup chopped bell pepper*
- ¾ *cup chopped celery*
- 2 *tablespoons Worcestershire sauce*
- 1 *tablespoon minced garlic*
- 1 *tablespoon salt*
- 1 *teaspoon hot sauce*
- ¼ *teaspoon white pepper*
- ¼ *teaspoon black pepper*
- ¼ *cup chopped green onions*
- ¼ *cup chopped parsley*
- 8 *cups cooked Toro Brand Rice*

DIRECTIONS

1. Rinse beans, being careful to remove any pieces of dirt or rock. Combine beans and beef broth in 8 to 10-quart pot. Bring to a boil.
2. Heat skillet, add **Bryan Smoked Sausage** and cook until brown. Drain on paper towels. Add sausage to beans.
3. Add remaining ingredients except green onions, parsley and cooked rice. Reduce heat to medium and simmer beans for 2½ hours. Make sure all of the water does not evaporate while simmering. If necessary add water to keep beans creamy.
4. Stir in green onions and parsley. Serve each portion of beans with a cup of cooked rice.

Yields 8 servings.

HAM AND ASPARAGUS QUICHE

Let me make this perfectly clear: "Real Men" do eat quiche.

- 5 *eggs*
- ½ *cup half-and-half cream*
- 1 *cup diced* ***Bryan Centerpiece Ham***
- ½ *cup shredded cheddar cheese*
- ¼ *cup minced green onions*
- ¼ *cup minced parsley*
- ½ *teaspoon salt*
- ¼ *teaspoon white pepper*
- 1 *(9-inch) unbaked pie shell*
- 10 *cooked asparagus spears*

DIRECTIONS

1. Preheat oven to 400°. In a bowl beat together eggs and half-and-half cream.
2. Stir in **Bryan Centerpiece Ham,** cheese, green onions, parsley, salt and pepper.
3. Pour mixture into pie shell and arrange asparagus in a circle over egg mixture.
4. Place in oven and bake for 40 minutes.
5. Allow to cool slightly before slicing.

Yields 4 entrée servings or 8 side servings.

ITALIAN RISOTTO

Risotto is one of my favorite dishes and, although I grew up in the rice capital of Louisiana, Crowley, I didn't discover this dish until my adult food-loving years.

1 tablespoon extra virgin olive oil
1 tablespoon unsalted butter
½ cup julienned shallots
1 tablespoon minced garlic
1 cup uncooked arborio rice
½ cup white wine
3 cups chicken broth, heated
2 teaspoons salt
¼ teaspoon black pepper
1 tablespoon dried basil leaves
1 tablespoon minced parsley

DIRECTIONS

1. In saucepot, heat olive oil and butter until butter is melted. Add shallots and garlic and cook for 5 minutes.
2. Lower fire and add arborio rice. Cook for 2 minutes, stirring constantly. Add white wine and cook until wine has been absorbed, stir constantly.
3. Add 1 cup of chicken broth and bring to a boil, lower fire, stir constantly until stock is absorbed.
4. Add another cup of broth and continue cooking and stirring until broth is absorbed.
5. Add remaining cup of broth, salt, pepper, basil and parsley, continuing to cook until last of broth is absorbed.
6. Remove from heat and serve with your favorite meat, seafood or poultry dish.

Yields 4 servings.

LOUISIANA DIRTY RICE DRESSING

This dish is referred to as "dirty rice" because of the color of the rice at the end of the cooking process. If you ask someone from Louisiana what kind of dressing they want on their salad, the answer just may be "rice!"

- 1 *pound ground pork*
- 1 *pound ground chuck*
- 2 *cups chopped onions, divided*
- 1 *cup chopped green bell pepper, divided*
- 1 *cup chopped celery, divided*
- 2 *tablespoons minced garlic*
- ½ *cup dark roux*
- 1 *tablespoon Tony Chachere's Creole Seasoning*
- 1 *teaspoon salt*
- 6 *cups beef broth*
- 2 *tablespoons Worcestershire sauce*
- 2 *bay leaves*
- 1 *teaspoon dried thyme leaves*
- 1 *teaspoon hot sauce*
- 1 *teaspoon black pepper*
- ½ *cup chopped green onions*
- ¼ *cup chopped parsley*
- 15 *packed cups cooked* **Toro Brand Rice**

DIRECTIONS

1. Heat a saucepot over medium heat. Add ground pork and chuck; cook until meat is browned.
2. Add 1 cup of onion, ½ cup bell pepper, ½ cup celery and garlic; cook for 2 minutes.
3. Add dark roux and cook for additional 5 minutes. Add Creole seasoning, salt, beef broth, Worcestershire sauce, bay leaves, thyme, hot sauce and black pepper.
4. Bring mixture to a boil, lower heat and simmer for 30 minutes.
5. Add remaining onion, celery and bell pepper; cover pot and simmer for additional 30 minutes.
6. Stir in green onions, parsley and cooked rice until thoroughly blended.

Yields approximately 20 servings.

ORZO PASTA JAMBALAYA

I substituted the rice-shaped pasta, orzo, for the more traditional ingredient of long grain rice. Be sure to add the cooked orzo pasta right at the end and simmer just until it is warm.

- 1 *pound orzo pasta*
- 5 *quarts water*
- 1 *tablespoon salt*
- 3 *tablespoons olive oil, divided*
- 1½ *pounds raw boneless chicken thigh meat*
- 1 *tablespoon Tony Chachere's Creole Seasoning*
- ½ *pound* **Bryan Smoked Sausage,** *sliced*
- ¼ *pound tasso ham, chopped*
- 2 *cups diced onion*
- 1 *cup diced celery*
- 1 *cup diced bell pepper*
- 1 *tablespoon minced garlic*
- 2 *(14-ounce) cans chicken broth*
- 1 *tablespoon unsalted butter, at room temperature*
- 1 *tablespoon flour*
- ¼ *cup thinly sliced green onions*
- ¼ *cup minced parsley*

DIRECTIONS

1. In large pot bring water and salt to a boil. Cook pasta according to package directions, drain, run cold water over pasta until cooled. Stir in 1 tablespoon olive oil until pasta is coated.
2. In another pot, heat remaining olive oil over medium heat; add chicken and Creole seasoning and cook for approximately 10 minutes, until chicken is completely brown.
3. Add **Bryan Smoked Sausage** and tasso. Brown an additional 5 minutes.
4. Stir in onion, celery, bell pepper and garlic and cook for another 10 minutes.
5. Add chicken broth and bring to boil. Lower fire and simmer for 30 minutes.
6. In a cup, stir together butter and flour to create a paste. Stir into chicken mixture and cook until slightly thickened, about 10 minutes.
7. Stir in cooked pasta, green onions and parsley until thoroughly blended.

Yields 8 to 10 servings.

PAN-FRIED EGGPLANT

To salt or not to salt the eggplant before frying, that is the question. I salt, end of story!

- 2 *cups water*
- 1 *teaspoon salt*
- 1 *large eggplant, peeled and sliced into medallions*
- 2 *cups cooking oil*
- 2 *cups flour*
- 1 *teaspoon Tony Chachere's Creole Seasoning, divided*
- 2 *eggs*
- 2 *cups milk*
- ½ *teaspoon hot sauce*
- 2 *cups Italian bread crumbs*

DIRECTIONS

1. In a bowl stir together salt and water. Place eggplant medallions in bowl and allow them to soak for 30 minutes. Drain eggplant, rinse with cold water and pat dry. Return to bowl.
2. Heat cooking oil to 325°.
3. In another bowl, season flour with ½ teaspoon Creole seasoning.
4. In another bowl, beat together eggs, milk, remaining Creole seasoning and hot sauce. In another bowl, place Italian bread crumbs.
5. Coat eggplant in seasoned flour, then dip into milk-egg mixture and then coat with bread crumbs, pressing firmly to make crumbs stick.
6. Place breaded eggplant in heated oil. Fry 2 to 3 minutes on each side until brown.

Yields approximately 10 fried eggplant medallions.

PASTA PRIMAVERA

One of the greatest culinary combinations known to man is heavy cream, Parmesan cheese and pasta. Exercise later!

- 2 *tablespoons olive oil*
- 1 *cup minced onion*
- 1 *tablespoon minced garlic*
- 1 *cup julienned green bell pepper*
- 1 *cup sliced celery*
- 1 *cup sliced mushrooms*
- 1 *cup sliced zucchini*
- 5 *cups heavy whipping cream*
- ½ *cup grated Parmesan cheese*
- 2 *tablespoons dried basil leaves*
- 2 *teaspoons dried oregano leaves*
- 1 *tablespoon Tony Chachere's Creole Seasoning*
- 1 *teaspoon hot sauce*
- 1 *bunch asparagus, cooked*
- 2 *cups of broccoli florets, cooked*
- 1 *pound of your favorite pasta, cooked according to package directions*

DIRECTIONS

1. In a large saucepot, heat olive oil over a medium heat. Add onions and garlic and cook for 5 minutes, being careful not to brown.
2. Stir in bell pepper, celery, mushrooms and zucchini; cook for 2 minutes.
3. Stir in heavy cream, Parmesan cheese, basil, oregano, Creole seasoning, hot sauce. Bring to a boil, lower fire and cook until cream starts to reduce, thicken slightly and form a sauce. Be careful not to reduce too much or cream could curdle.
4. Stir in asparagus and broccoli; simmer until heated.
5. Ladle primavera over pasta.

Yields 8 servings.

Cajun scarecrow

PECAN RICE

Roasting the pecans in a 350° oven for 10 minutes prior to adding them to the dish will intensify the flavor of the pecans and this recipe.

- 3 *tablespoons unsalted butter*
- 1 *cup chopped onion*
- ½ *cup chopped green bell pepper*
- ½ *cup chopped celery*
- 1 *tablespoon minced garlic*
- 1 *tablespoon Tony Chachere's Creole Seasoning*
- 1 *cup uncooked* ***Toro Brand Rice***
- 4 *cups beef broth, divided*
- 1 *cup roasted pecans*
- 1 *teaspoon hot sauce*
- ¼ *cup minced parsley*
- ½ *cup chopped green onions*

DIRECTIONS

1. In saucepot, heat butter over medium heat. Add onion, bell pepper and celery and cook for 5 minutes. Turn heat up to high.
2. Add garlic and cook for an additional 2 minutes. Add Creole seasoning and **Toro Brand Rice** and cook for an additional minute.
3. Stir in 3½ cups of beef broth, bring to boil and cook for 5 minutes. Lower heat to medium and stir in remaining ½ cup broth, pecans and hot sauce. Cover and cook for 10 minutes.
4. Remove from heat and stir in parsley and green onions.

Yields 6 servings.

PORK AND EGGPLANT RICE DRESSING

When smothering down the eggplant as we do in this dish, it is not necessary to salt the eggplant prior to cooking.

1 pound fresh pork sausage
1½ cups chopped onion
½ cup chopped bell pepper
½ cup chopped celery
1 tablespoon minced garlic
⅓ cup dark roux
2 (14-ounce) cans beef broth
1 tablespoon Tony Chachere's Creole Seasoning
1 large eggplant, peeled and diced (approximately 6 cups)
*6 cups cooked **Toro Brand Rice***
¼ cup chopped green onion
¼ cup chopped parsley

DIRECTIONS

1. Squeeze fresh pork sausage out of casing into saucepot and cook over medium heat until completely browned.
2. Stir in onion, bell pepper and celery and cook for 5 minutes. Add garlic and roux. Cook for additional 5 minutes, stirring until roux blends with onion mixture.
3. Add beef broth and Creole seasoning. Bring to boil, cover, reduce heat to simmer and cook for 20 minutes.
4. Uncover and add eggplant. Bring back to boil, reduce heat, cover and simmer for additional 30 minutes. Uncover and simmer for 5 minutes
5. Stir in cooked **Toro Brand Rice,** green onions and parsley.

Yields 8 to 10 servings.

PRALINE SWEET POTATO CASSEROLE

I have seen people eat this dish for dessert!

1 (14-ounce) can mashed sweet potato
2 eggs
1 teaspoon cinnamon
1 teaspoon Tony Chachere's Creole Seasoning
1 teaspoon hot sauce
½ teaspoon baking powder
½ teaspoon salt
¼ teaspoon nutmeg

PRALINE TOPPING

½ cup brown sugar, packed
½ cup white sugar
2 tablespoons unsalted butter
1½ cups pecan pieces

DIRECTIONS

1. Preheat oven to 350°.
2. In a bowl, beat together sweet potato, eggs, cinnamon, Creole seasoning, hot sauce, baking powder, salt and nutmeg.
3. Place sweet potato mixture in casserole dish and bake for 30 minutes. Remove from oven and top with praline topping; bake for an additional 5 minutes.
4. Remove from oven and cool slightly before serving.

PRALINE TOPPING

1. Combine all ingredients except pecan pieces in a saucepot; bring to boil and lower fire. Cook to 234° on candy thermometer.
2. Stir in pecans and pour over casserole.

Yields 8 side servings.

Debbie's grandfather's table

SEAFOOD BROCCOLI CASSEROLE

The crunchy cheesy topping is a nice contrast to the creamy consistency of the casserole.

- 1 *tablespoon olive oil*
- ½ *cup chopped onion*
- ¼ *cup chopped celery*
- ¼ *cup chopped bell pepper*
- 1 *tablespoon chopped garlic*
- 1 *(10-ounce) can* ***Campbell's Cream of Broccoli Soup***
- 1 *(10-ounce) can* ***Campbell's Cream of Shrimp Soup***
- ½ *cup heavy cream*
- 2 *(10-ounce) packages frozen broccoli cuts*
- ½ *pound shrimp, peeled*
- ½ *pound crabmeat*
- 1 *teaspoon Tony Chachere's Creole Seasoning*
- 1 *teaspoon dried thyme leaves*
- 1 *teaspoon salt*
- ¼ *teaspoon white pepper*
- 1 *cup grated cheddar cheese*
- ¼ *cup Italian bread crumbs*

DIRECTIONS

1. Preheat oven to 350°.
2. In a saucepot, heat olive oil over a medium heat. Add onions, celery and bell pepper, and cook for 5 minutes. Add garlic and cook for an additional minute.
3. Add **Campbell's Cream of Broccoli Soup, Campbell's Cream of Shrimp Soup** and heavy cream. Cook, stirring, until soups become smooth.
4. Stir in broccoli, shrimp, crabmeat, Creole seasoning, thyme, salt and pepper. Reduce heat and simmer for 10 minutes. Pour mixture into casserole dish.
5. In a bowl toss together cheese and bread crumbs and top casserole with mixture. Bake for 20 minutes.
6. Cool slightly before serving.

Yields 12 servings.

SEASONED RICE PILAF

This is an excellent change of pace from plain white rice.

2 tablespoons unsalted butter
½ cup chopped onion
½ cup shredded carrots
¼ cup chopped bell pepper
¼ cup chopped celery
1 tablespoon minced garlic
*1 cup **Toro Brand Rice***
2 cups chicken broth
1 teaspoon hot sauce
¼ cup minced green onions
¼ cup minced parsley

DIRECTIONS

1. In a saucepot, heat butter over medium heat. Add onions, carrots, bell pepper and celery, and cook for 5 minutes.
2. Stir in garlic and cook an additional minute. Stir in **Toro Brand Rice** and cook for 2 minutes while stirring.
3. Stir in chicken broth and hot sauce, bring to a boil and cook 5 minutes.
4. Lower fire, cover pot and cook an additional 10 minutes.
5. Stir in green onions and parsley.

Yields 4 side servings.

SMOKED ANDOUILLE RICE CAKES

Although these moist, smoky morsels can be served as a side dish, I like to serve them on a bed of Creole red beans or crawfish étouffée as an entrée.

- 3 *tablespoons cooking oil*
- ¾ *cup chopped onion*
- ½ *cup chopped green bell pepper*
- ¼ *cup chopped celery*
- ½ *pound minced andouille sausage*
- 3 *tablespoons flour*
- 3 *cups chicken broth*
- 4 *tablespoons Tony Chachere's Creole Seasoning, divided*
- 1 *teaspoon hot sauce*
- 3 *cups cooked* ***Toro Brand Rice***
- 2 *tablespoons minced green onion*
- 1 *tablespoon minced parsley*
- 2 *cups flour*
- *Cooking oil for pan sautéing*

DIRECTIONS

1. In large skillet heat oil. Cook onion, celery, bell pepper and andouille sausage until vegetables start to brown.
2. Stir in flour. Whisk in chicken broth, 2 tablespoons Creole seasoning and hot sauce. Bring to boil.
3. Reduce heat and simmer for 5 minutes until sauce thickens. Remove from heat.
4. Cool the sauce for 20 minutes, stirring occasionally.
5. Stir in **Toro Brand Rice,** green onions and parsley. Refrigerate rice mixture until completely chilled. At this point, it may be refrigerated overnight or formed into cakes and cooked.
6. Form rice cake mixture into patty using a large ice cream scoop (#12 scoop) or about ⅓ cup each.
7. Season flour with remaining Creole seasoning. Lightly dust patty in seasoned flour and pan sauté in heated oil for several minutes on both sides to ensure that the middle of the patty is thoroughly heated.
8. Serve alone or on a bed of red beans or crawfish étouffée.

Yields approximately 14 rice cakes.

SMOKED SAUSAGE, SHRIMP AND EGGPLANT

By adding a small amount of sugar to this dish, you will cut any bitterness you may get from the eggplant or tomatoes.

- ½ *pound* ***Bryan Smoked Sausage,*** *diced*
- 2 *cups diced onion*
- 1 *cup diced bell pepper*
- 1 *cup diced celery*
- 1 *tablespoon minced garlic*
- ¼ *cup dark roux*
- 1 *(14-ounce) can beef broth*
- 1 *(10-ounce) can diced Rotel Tomatoes*
- 3 *pounds eggplant, peeled, cubed medium*
- 2 *tablespoons sugar*
- 1 *teaspoon dried oregano leaves*
- 1 *teaspoon dried thyme leaves*
- ½ *teaspoon dried basil leaves*
- 3 *bay leaves*
- 1 *pound medium shrimp, peeled*
- ¼ *cup minced green onions*
- 2 *tablespoons minced parsley*

DIRECTIONS

1. Heat large pot over medium heat and cook **Bryan Smoked Sausage** until browned. Stir in onion, bell pepper and celery and cook for 5 minutes. Add garlic and roux and cook for 5 minutes until roux dissolves.

2. Stir in beef broth, Rotel Tomatoes, eggplant, sugar, oregano, thyme, basil and bay leaves. Cover pot and simmer for 30 minutes; uncover and simmer for an additional 5 minutes.

3. Remove bay leaves and discard; add shrimp and cook covered for five minutes. Uncover pot and simmer for additional 5 minutes.

4. Stir in green onions and parsley.

Yields 12 side dishes or 6 entrées.

Vegetarian Gator

SMOKED WHITE BEANS

This recipe is a unique way to utilize the water pan of your smoker. As the meat of your choice is cooking above, it is adding flavor to the beans that are cooking below in the water pan.

1 pound dried navy beans
5 (10½-ounce) cans chicken broth
1 cup chopped onion
½ cup chopped celery
½ cup chopped green bell pepper
2 tablespoons minced garlic
1 teaspoon Tony Chachere's Creole Seasoning
1 tablespoon hot sauce
2 bay leaves

DIRECTIONS

1. Clean and rinse beans, removing any dirt and rocks found.
2. Combine all ingredients in the water pan of your smoker.*
3. Smoke beans for 2 hours.
4. Remove from smoker; pour beans into a pot, place on stove and simmer for 1 hour and 15 minutes, stirring occasionally. If liquid runs low, add enough water to maintain a creamy consistency.

Yields 1 gallon cooked beans.

**If you don't have a smoker the beans can be placed in a pot and cooked on your BBQ pit with the coals and wood off to the side. Use a pot that you are not worried about getting discolored from the smoke.*

TASSO AND GREEN BEAN CASSEROLE

Tasso is smoked meat, usually pork that is marinated in a seasoned brine solution then smoked over a southern hardwood. It is used to add flavor to a wide variety of dishes and works particularly well with vegetables — in this case green beans.

- 2 *tablespoons unsalted butter*
- 1 *cup tasso*
- ½ *cup chopped onion*
- ¼ *cup chopped celery*
- ¼ *cup chopped bell pepper*
- 1 *tablespoon minced garlic*
- 1 *(10-ounce) can of* ***Campbell's Cream of Mushroom Soup***
- 1 *pound frozen cut green beans*
- ¼ *cup heavy cream or milk*
- 1 *teaspoon Tony Chachere's Creole Seasoning*
- 1 *teaspoon hot sauce*
- ¼ *teaspoon white pepper*
- ½ *cup roasted sliced almonds*

DIRECTIONS

1. In skillet heat butter. Cook tasso, onion, celery, bell pepper and garlic for two minutes.
2. Add **Campbell's Cream of Mushroom Soup,** cut green beans, heavy cream, Creole seasoning, hot sauce and white pepper. Simmer for 10 minutes or until the green beans are cooked.
3. Heat oven to 350° and roast almonds for 10 minutes. Top each serving of casserole with small amount of roasted almonds.

Yields 8 side servings.

DESSERTS

DESSERTS

Acadian Bread Pudding with Roasted Rum Pecan Sauce
Bananas Foster
Cherries Jubilee
Chocolate Amandine Sauce
Chunky Chocolate Walnut
  Bread Pudding with Kahlúa Sauce
Crème Brûlée
Praline Cheesecake with Praline Sauce
Profiteroles with Vanilla Ice Cream
Pumpkin-Chocolate Cheesecake with Caramel Sauce
Southern Pecan Pie
Summer Blueberry Cobbler
White Chocolate Bread Pudding with White Chocolate
  Kahlúa Sauce

ACADIAN BREAD PUDDING WITH ROASTED PECAN RUM SAUCE

This bread pudding I have developed is short on sugar and instead is spiced mildly with nutmeg and cinnamon. The crunch from the pecans in the rum sauce is a nice balance to the creaminess of this traditional custard-based dessert.

BREAD PUDDING

6 cups cubed day-old French bread
4 eggs
1½ cups granulated sugar
3½ cups milk
1 teaspoon ground cinnamon
½ teaspoon ground nutmeg
1 stick unsalted butter, melted

ROASTED PECAN RUM SAUCE

½ cup medium pecan pieces
2 cups heavy whipping cream
4 tablespoons granulated sugar
4 tablespoons light rum
1 tablespoon cornstarch
1 tablespoon water
1 teaspoon orange zest

DIRECTIONS

BREAD PUDDING

1. Preheat oven to 350°. Place French bread in an 12 x 8-inch baking dish.
2. In a bowl, beat together eggs and sugar for 3 minutes. Add the milk, nutmeg, cinnamon, and butter and continue to beat until blended.
3. Pour the milk-egg mixture over the French bread. Let sit for 1 hour, stirring occasionally, until bread soaks up custard.
4. Place in oven and bake for 30 minutes, then lower temperature to 300° and bake an additional 30 minutes or until puffy and brown.
5. Top each serving of bread pudding with approximately 3 tablespoons of Roasted Pecan Rum Sauce.

ROASTED PECAN RUM SAUCE

1. Place the pecans on a baking sheet and roast in a 350° oven for 5 minutes. Remove from oven and set aside.
2. In a saucepot, place heavy cream and bring to simmer over a medium heat. Add the sugar and rum; whisk the mixture until the sugar is dissolved.
3. In a cup, dissolve cornstarch in water. Stir this mixture into the heated cream, and stir for 1 minute until the mixture is thickened.
4. Stir in the orange zest and roasted pecans.

Yields 12 servings.

BANANAS FOSTER

A classic New Orleans dessert that is loved by young and old. If you are serving it to the kids you may eliminate the rum.

¼ cup orange juice
½ stick unsalted butter
4 tablespoons brown sugar
1 tablespoon lemon juice
1 teaspoon cinnamon
½ teaspoon nutmeg
⅓ cup light rum
4 semi-ripe bananas, peeled and quartered
4 scoops vanilla ice cream

DIRECTIONS

1. In a skillet, place orange juice, butter, brown sugar, lemon juice, cinnamon and nutmeg; stir over medium heat until butter is melted and sugar is dissolved.
2. Add rum, being careful as it may flame up. Simmer for 1 minute.
3. Add bananas and simmer for additional 2 minutes.
4. Serve over vanilla ice cream.

Yields 4 servings.

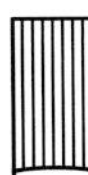

CHERRIES JUBILEE

Cherries Jubilee is a classic dessert that is an impressive end to any great dinner.

- 6 *tablespoons unsalted butter*
- 4 *tablespoons brown sugar*
- ¼ *cup maraschino cherry syrup*
- 2 *tablespoons cane syrup*
- ½ *cup Grand Marnier*
- ¼ *teaspoon cinnamon*
- 1 *tablespoon cornstarch*
- 1 *tablespoon water*
- 1 *(16-ounce) can red pitted cherries in water, drained; reserve ½ cup of liquid*
- 4 *scoops vanilla ice cream*

DIRECTIONS

1. In small saucepot heat butter and brown sugar until sugar is dissolved.
2. Add maraschino syrup and cane syrup and bring to a simmer.
3. Add Grand Marnier, being careful as it could flame up. Stir in cinnamon.
4. In a cup, dissolve cornstarch in water. Stir this mixture into sauce, and stir for 1 minute until mixture is thickened.
5. Add cherries and reserved cherry liquid and simmer for additional minute. Serve over ice cream.

Yields 4 servings.

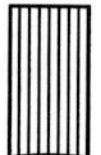

CHOCOLATE AMANDINE SAUCE

The orange accent from the Grand Marnier and the orange zest adds that little zing to the dish that will keep your guests guessing.

½ cup sliced almonds
½ cup semisweet chocolate chips
¼ cup strong coffee
1 tablespoon orange zest
2 tablespoons unsalted butter
2 tablespoons heavy whipping cream
*2 tablespoon Grand Marnier (optional)**
4 scoops vanilla ice cream

DIRECTIONS

1. Preheat oven to 350°. Place almonds on a cookie sheet and roast in oven until slightly browned, approximately 10 minutes. Set aside to cool.
2. In a double boiler combine chocolate, coffee and orange zest. Simmer over low heat until chocolate melts, stirring until chocolate is smooth.
3. Remove from heat and stir in butter until melted.
4. Stir in heavy cream and Grand Marnier until completely incorporated. Fold in roasted almonds.
5. Top each serving of vanilla ice cream with chocolate amandine sauce.

Yields 1 cup of sauce.

**You may substitute water for the Grand Marnier.*

CHUNKY CHOCOLATE WALNUT BREAD PUDDING WITH KAHLÚA SAUCE

This is another version of the traditional custard pudding that people have come to associate with South Louisiana.

- 6 *cups cubed day-old French bread*
- 1 *cup granulated sugar*
- 4 *large eggs*
- 3 *cups milk*
- 1 *tablespoon pure vanilla extract*
- 1¼ *cups* ***Nestlé® Tollhouse® Semi-Sweet Chocolate Morsels***
- 1 *cup medium walnut pieces*

KAHLÚA CREAM SAUCE

- 2 *cups heavy cream*
- ½ *cup sugar*
- ½ *cup Kahlúa*
- 1 *tablespoon cornstarch*
- 1 *tablespoon water*

DIRECTIONS

1. Preheat oven to 350°. Place French bread in 12 x 8-inch baking dish.
2. In bowl, combine sugar and eggs and whip for 3 minutes.
3. Add milk and vanilla and continue to whip until blended. Pour the milk-egg mixture over French bread. Allow French bread to soak for 1 hour, stirring occasionally, until bread soaks up milk-egg mixture.
4. Stir in **Nestlé® Tollhouse® Semi-Sweet Chocolate Morsels** and walnut pieces until evenly dispersed throughout the pudding.
5. Place in oven and bake for 30 minutes. Lower temperature to 300° and bake an additional 30 minutes or until puffy and brown.

SAUCE

1. In a small saucepot, stir together heavy cream, sugar and Kahlúa over medium heat until mixture come to a boil. Lower fire and simmer for a minute.
2. In a cup, dissolve cornstarch in water. Whip into heated cream mixture and stir until thickened.
3. Top individual servings of bread pudding with 3 tablespoons sauce.

Yields approximately 12 servings.

mayonnaise

CRÈME BRÛLÉE

This has always been one of my favorite desserts. The trick is making sure the custard cooks long enough to set up when it chills so that it's dense yet creamy. The small blowtorch I use to caramelize the sugar topping can be found in any hardware store.

3 egg yolks
1 cup granulated sugar
2 cups heavy cream
1 tablespoon vanilla extract
4 tablespoons brown sugar

DIRECTIONS

1. Preheat oven to 250°.
2. In a bowl, combine egg yolks and granulated sugar and beat together until completely creamed and lemon colored. Set aside.
3. In a saucepot stir together heavy cream and vanilla extract over medium heat until it begins to simmer. Remove from heat.
4. Slowly pour cream into egg mixture, whipping until cream is incorporated. Strain mixture through a fine sieve.
5. Divide mixture into four (6-ounce) enamel baking dishes. Place dishes in a roasting pan. Carefully fill pan with water to a depth of 1 inch.
6. Bake uncovered for 50 minutes.
7. Remove from pan and place in refrigerator for 6 hours until completely cool.
8. Before serving, spread 1 tablespoon of brown sugar evenly over each baking dish. Heat sugar until it caramelizes by placing custards under a preheated broiler or heating with a blowtorch.

Yields 4 servings.

PRALINE CHEESECAKE WITH PRALINE SAUCE

This is a pretty basic cheesecake recipe; what makes it so special is the praline sauce I use as a topping. It's so good you may be tempted to eat the sauce by itself.

GRAHAM CRACKER CRUST

1 *cup graham cracker crumbs*
¼ *cup sugar*
3 *tablespoon unsalted butter, melted*

FILLING

3 *pounds cream cheese, at room temperature*
1 *cup sugar*
½ *cup light brown sugar*
8 *eggs*
4 *tablespoons flour*
2 *tablespoons pure vanilla extract*
1 *cup medium pecan pieces*

PRALINE SAUCE

1 *egg, beaten*
1 *cup sugar*
1 *cup light corn syrup*
½ *cup brown sugar*
½ *cup butter*
½ *cup medium pecan pieces*

DIRECTIONS

CRUST

1. Mix ingredients together in a mixing bowl until moist and crumbly.
2. Using the back of a large spoon, press crumb mixture firmly on bottom of 9-inch springform pan. Set aside.

FILLING

1. Preheat oven to 325°. Place softened cream cheese in a bowl, add sugars and beat until light and fluffy.
2. Beat eggs, one at a time into cream cheese mixture until completely incorporated.
3. Add remaining ingredients and beat until blended. Pour batter into springform pan lined with graham cracker crust.
4. Bake for 1 hour. Turn off oven and leave in oven for 1 hour. Remove from oven and allow to chill completely in the refrigerator before slicing and topping with small amount of praline sauce.

PRALINE SAUCE

1. In a saucepot stir together all ingredients except pecan pieces. Bring to boil, lower heat and simmer for 2 minutes
2. Add pecan pieces and allow sauce to cool to room temperature before topping slice of cheesecake. You may garnish cheesecake slices with whipped cream and fresh fruit.

Yields 12 to 16 slices.

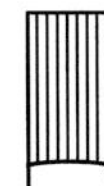

PROFITEROLES WITH VANILLA ICE CREAM

There are not many people who don't enjoy these airy puffs of dough, especially when you combine them with chocolate and ice cream.

½ *cup water*
¼ *stick unsalted butter*
½ *cup flour*
¼ *teaspoon salt*
2 *large eggs*
Nonstick cooking spray
16 *small scoops vanilla ice cream*
Chocolate Amandine Sauce *recipe, page 127*

DIRECTIONS

1. In a saucepot bring water to a boil. Lower fire, add butter and simmer until butter melts.
2. Stir in flour and salt with a spoon until mixture is smooth. Cook until mixture pulls away from the sides of saucepot, forming a ball. This should take about 1 minute.
3. Remove from fire and stir in slightly beaten eggs, one at a time, until completely incorporated. Cover and let stand until completely cooled.
4. Preheat oven to 425°. Spray baking sheet with nonstick cooking spray and drop dough by the heaping tablespoonful onto the pan. You should get 16 profiteroles. Bake for 18 to 20 minutes or until golden brown.
5. Turn off oven.
6. Remove pan from oven and pierce each profiterole with knife, allowing the steam to escape. Place back in oven, leave oven door open slightly, and allow profiteroles to dry for 10 minutes. Remove the pan from oven and allow profiteroles to completely cool.
7. Split the puffs in half and place small scoop of vanilla ice cream on bottom of profiterole then top with other half. Place 2 filled profiteroles on a plate and drizzle with **Chocolate Amandine Sauce.**

Yields 8 servings.

PUMPKIN-CHOCOLATE CHEESECAKE WITH CARAMEL SAUCE

This recipe has become a tradition for my family around the holidays!

GRAHAM CRACKER CRUST

1 *cup graham cracker crumbs*
¼ *cup granulated sugar*
3 *tablespoons butter, melted*

FILLING

3 *pounds cream cheese, at room temperature*
1 *cup granulated sugar*
8 *eggs*
1 *(15-ounce) can* ***Libby's® 100% Pure Pumpkin***
4 *tablespoons flour*
2 *tablespoons vanilla extract*
1 *cup* ***Nestlé® Tollhouse® Semi-Sweet Chocolate Morsels***

CARAMEL SAUCE

½ *cup granulated sugar*
2 *teaspoons water*
1 *cup* ***Nestlé® Carnation® Evaporated Milk***
2 *egg yolks*

DIRECTIONS

GRAHAM CRACKER CRUST

1. In a bowl, mix ingredients together until moist and crumbly.
2. Using the back of a large spoon, press crumb mixture firmly on the bottom of a 9-inch springform pan. Set aside.

FILLING

1. Preheat oven to 325°.
2. Place softened cream cheese in a bowl. Add sugar and beat until light and fluffy.
3. Beat eggs, one at a time, into cream cheese mixture until completely incorporated.
4. Beat in **Libby's® 100% Pure Pumpkin,** flour and vanilla until completely incorporated.
5. Fold in **Nestlé®® Tollhouse® Semi-Sweet Chocolate Morsels.**
6. Pour pumpkin-chocolate cheesecake mixture into springform pan.
7. Bake for 1 hour. Turn oven off and leave in oven for 1 hour.
8. Remove from oven and allow to chill completely in refrigerator before slicing.
9. Top each slice of cheesecake with small amount of Caramel Sauce.

CARAMEL SAUCE

1. In saucepot heat sugar until melted. Stir in water and continue to cook until sugar is browned but not burned.
2. Add **Nestlé® Carnation® Evaporated Milk** and egg yolks and continue to cook until sauce is thick enough to coat back of spoon.
3. Serve warm over cheesecake slices.

Yields 12 to 16 slices.

SOUTHERN PECAN PIE

I find most pecan pies entirely too sweet, but not this recipe. I like to serve it warm with either whipped cream or a scoop of vanilla ice cream.

- ¾ *cup vegetable shortening, chilled*
- 1 *cup flour*
- ¾ *cup plus 1 teaspoon light corn syrup, divided*
- ¼ *cup chilled water*
- 1½ *cups medium pecan pieces*
- 4 *eggs*
- ¾ *cup sugar*
- 1 *tablespoon unsalted butter, melted*
- 1 *teaspoon pure vanilla extract*

DIRECTIONS

1. Preheat oven to 325°. In a large bowl cut shortening into flour until mixture resembles small peas. In another bowl combine chilled water, 1 teaspoon of corn syrup and salt, stirring until salt has dissolved. Add flour mixture and stir until blended. Form dough into ball, wrap in plastic wrap and chill in the refrigerator for 1 hour.

2. On a lightly floured surface roll out dough to ¼ inch thickness. Line 9-inch deep-dish pie pan with dough. Spread pecan pieces over dough. Place on sheet pan and set aside.

3. In a bowl beat together eggs and sugar for 1 minute. Add remaining ingredients and beat until mixture is smooth.

4. Pour filling into pie shell. Place in oven and bake for 50 minutes or until filling is firm. Remove from oven and allow pie to sit for 2 hours before slicing.

Yields 8 to 10 slices.

SUMMER BLUEBERRY COBBLER

This is a great dessert to bake with the kids. You can substitute their favorite fruit if you wish. Serve up some homemade ice cream on the side; kids love to crank an old-time ice cream maker.

1¼ cups flour
5 tablespoons unsalted butter
2 teaspoons baking powder
½ teaspoon salt
8 tablespoons half-and-half cream
4 tablespoons cane syrup
3½ cups fresh or frozen blueberries
¾ cup granulated sugar
2 tablespoons cornstarch
2 tablespoons orange juice
1 tablespoon orange zest
½ teaspoon cinnamon
10 cups vanilla ice cream

DIRECTIONS

1. Preheat oven to 375°.
2. In a bowl, combine flour, butter, baking powder and salt. Using a fork or your fingers, knead flour and butter mixture until it resembles corn meal.
3. Add half-and-half cream and cane syrup. Continue to mix until a soft dough is formed. Place on floured board and roll into flat crust, 8 x 10 x 1-inch. Set aside.
4. Combine berries, sugar, cornstarch, orange juice, orange zest and cinnamon in a bowl and stir until blended.
5. Place filling in 8 x 10-inch casserole dish and spread dough over top. Press dough down around edges to seal filling. Pierce holes in the top of dough with a fork to release any steam from cobbler during baking process.
6. Bake for 30 minutes until dough is evenly browned. Cool for 5 minutes. Serve each portion of cobbler with a scoop of vanilla ice cream.

Yields 10 servings.

WHITE CHOCOLATE BREAD PUDDING WITH WHITE CHOCOLATE KAHLÚA SAUCE

This is my version of a recipe that was first created by the Palace Café, a great New Orleans restaurant.

- 6 *cups cubed day-old French bread*
- 4 *cups heavy cream*
- 1 *cup milk*
- ½ *cup sugar*
- ½ *pound white chocolate, coarsely chopped*
- 2 *whole eggs*
- 5 *egg yolks*

SAUCE

- 2 *cup heavy cream*
- ¼ *cup Kahlúa*
- ½ *pound white chocolate, coarsely chopped*

DIRECTIONS

1. Preheat oven to 350°.
2. Place cubed French bread in large casserole dish and toast in oven for 15 minutes. Remove from oven and set aside.
3. Place heavy cream, milk and sugar in a saucepot and heat just to the boiling point. Remove from stove, add white chocolate and stir until chocolate has melted.
4. Whip together whole eggs and egg yolks. Add to cream and whip until incorporated. Pour mixture over toasted French bread and allow to soak for 1 hour, stirring occasionally until French bread has soaked up custard. Cover and place in oven; bake 35 minutes. Uncover and bake for an additional 15 minutes.
5. Top each serving of bread pudding with 3 tablespoons of Kahlúa Sauce.

KAHLÚA SAUCE

1. Bring heavy cream and Kahlúa to a simmer in a small saucepot and cook for 3 minutes.
2. Remove from stove and add chocolate. Stir until chocolate is melted and sauce is smooth.

Yields 12 servings.

Medric Martin's store in Franklin

LAGNIAPPE

LAGNIAPPE

Banana Walnut Pancakes with Strawberry Butter
Chocolate Peanut Butter Pecan Cookies
Chocolate Pecan Pralines
Fig Nut Bread
Ham and Cheese Pancakes
Herb Buttermilk Drop Biscuits
New Orleans Muffuletta
Pain Perdu
Pasta Macque Choux
Pecan Pralines
Creole Sausage Bread Loaves
Sweet and Spicy Pecans

CONTENTS

Acknowledgments

YOU DON'T DO ANYTHING FOR TWENTY YEARS WITHOUT MEETING A FEW people along the way who share knowledge and friendship. People who inspire you to be successful at whatever profession you pursue. These are just a few of those people.

- My entire family especially my nephew B. J. Comeaux, thank you for believing in me.
- Charlie & Dell Goodson, Ken Veron and Richard Compton for giving me my first Executive Chef's job in spite of my youth and inexperience.
- The two chefs who taught me the most along the way, Tom Thompson and Leopold Langoria.
- All the restaurant managers, sous chefs, line cooks, dishwashers and wait staff that I have had the pleasure to work with over the course of my career; the Chef is only as good as the staff.
- The staff and management of KLFY TV-10 especially Carol Behrens, Joe Varholy, Mike Barras, Maria Placer, Sherman Richard and the entire production staff.
- The staff and management of Church Point Wholesale and the Lagniappe Food Stores especially Linda Gail Minsky, Jack Casanova, Penny Hartwick, Tom Brown and Keith Joubert.
- All the sponsors of "The Chef" program.
- My photographer Debbie Fleming Caffery, who helped me to recognize the artistry of what I do and whose art continues to inspire me daily.
- My editors Linda Gail Minsky, Alice Landry & Tim Basden, Robert Wolf, Shereen Minville and especially Lauren Pettingill, Sharon Arms Doucet and Judy Johnson.
- My graphic designer, Megan Barra, whose talent is unparalleled.
- Katrina Huggs Pardue for her assistance in writing the "About The Chef" and "Notes from The Chef" sections.
- Freddie Strange, Connie Morgan and the staff of the Wimmer Companies.
- And last but certainly not least, Dell Hains, Chef Wayne Jean, Rick and Elaine Guidry, Mike Landry, Greg Hahn, Michael Doucet, Randy Arceneaux, Jim Olivier, Sonny Landreth, Janet Mould, Mary Margaret Camalo, Dickie & Cynthia Breaux, Kerry Boutte, Marcelle Bienvenu, Chris Lee, Michelle Vallot, Tommy Simmons, Corinne Cook, Chef Paul Prudhomme, Joe Cahn, David Alpha, Dennis Paul Williams, Nathan Williams, Jr. & son, Chuckie Williams, Medric Martin, Jeff & Bonnie Venable, The Guidry Brothers, Christine Balfa and Sheryl Cormier.

Foreword

Chef Patrick Mould's Louisiana Cooking

WHO WOULD THINK THAT A RED-HAIRED, FRECKLED-FACE, NICE GUY FROM Acadia Parish could turn out to be someone who knew exactly how to cook and season the food of this bountiful area of Acadiana?

Pat began his association with KLFY TV-10 as a guest chef on various shows. It wasn't long before he had his own segment, "Louisiana Cooking," within the five o'clock show we call Newscan 10. The audience warmed to his cooking but also enjoyed his forays into the backwoods and swamps of South Louisiana. He showed us how and where the elements of his dishes came about, from wild game to a fancy feast.

Pat went on to host the popular segment "The Chef," which airs during the shows "Passe Partout" and "Meet Your Neighbor." Each week he shows us how to create the dishes that make him a unique master of the culinary arts. He shares his secrets with the audience and now, with this book, is making those secrets available to anyone who wants to know the basics and more of Louisiana cooking. Pat is always able to satisfy the palates of those lucky enough to sit at his table.

Not only is he well-liked because of his warm personality, he has also established himself as a true son of Acadiana with magic in his black pot and ladle. The relationship between Chef Patrick Mould and KLFY is very much like the roux that is the basis for many of Pat's dishes–it is warm, substantive and the foundation for something wonderful. In short, it's the kind of friendship one finds in Acadiana.

It's a pleasure to introduce to you the man and his culinary talents. You'll like the first, enjoy the second and spread the joy of Cajun cooking.

Maria Placer
Vice President of News KLFY-TV 10

About the Chef

WHEN THE POPULARITY OF Cajun food exploded on the national scene in the mid-1980s, Americans soon became familiar with dishes like Crawfish Etouffée and Seafood Gumbo. But while this traditional Cajun fare was becoming hot nationally and internationally, back home in the eight-parish region of Acadiana where Cajun food was born, many chefs hardly took note. They were too busy creating a renaissance of Cajun cooking.

Patrick Mould remembers working as the executive chef at the highly rated Café Vermilionville in Lafayette, Louisiana, in the early '80's when no one was sautéing, grilling or doing sauces over fish. But by the mid '80's—about the same time traditional Cajun food had become a fad across the country—a small group of progressive young chefs that included Mould took on the mission of elevating Cajun cuisine to new heights.

These chefs knew they could do more than just a good gumbo or etouffée. Their challenge was to broaden their horizons, but at the same time stay close to their culinary roots and be true to their Cajun heritage.

To expand upon their talents, Mould and his peers drew on what chefs like Larry Forgione and Wolfgang Puck were doing in other parts of the country. For inspiration closer to home they looked to Cajun food guru Paul Prudhomme. It was this combination of utilizing indigenous ingredients with other regional cooking techniques and foods that sparked the culinary renaissance in Acadiana. The new cuisine being cranked out was dubbed "Nouvelle Cajun Cuisine."

At the forefront of this culinary movement was Patrick Mould. In 1984, he won Best of Show for his "Redfish Louisiane" at the Acadiana Culinary Classic. The prestigious culinary competition and event, held annually in Lafayette, is sponsored by the Acadiana Chapter of the American

Culinary Federation. By adding ingredients such as bell pepper, shallots, white wine and a rich shrimp stock, Chef Mould put a little twist on a basic crawfish sauce that had its roots in the basic Crawfish Etouffée recipe of his ancestors. Winning Best of Show was not only a great honor for the chef, but more importantly, Mould's award christened a new era of Nouvelle Cajun Cuisine.

Since 1984, Nouvelle Cajun Cuisine has developed even further. Paralleling its development has been the rise of Patrick Mould's professional status. Born in Paris, France, and raised in the Cajun town of Crowley, Louisiana, today he is one of the most renowned chefs in the South.

When it comes to Cajun cuisine, Chef Patrick Mould is a maverick and his unique way of stirring the pot has won him enough gold and silver medals to weight him down. He has also accumulated a menu-full of titles, including President of the Acadiana Chapter of the American Culinary Federation from 1985 to 1987 and the Chapter's coveted "Chef of the Year" award in 1987. The Federation is a national organization of chefs and cooks that is recognized as the authority of food in America.

A Louisiana celebrity, Chef Patrick Mould has been featured in such magazines as *Southern Living, Town & Country, The Chicago Tribune Magazine* and *Louisiana Life*. He has also appeared in several culinary publications, including *Cook's* and *The Best of Food & Wine's 1988 Collection.* As comfortable in front of the camera as he is behind the stove, Mould has appeared on several nationally syndicated cooking shows. He has also served as chef host for the popular television show, "Cajun

Country USA" and produced Cajun cuisine segments for the BBC. He has even produced a cooking segment, "Louisiana Cooking," that aired during news programming on CBS-affiliate KLFY TV-10 in a culturally rich, field-reporting style. A celebrity's chef as well as a celebrity chef, Patrick Mould's career highlights include preparing dinner hosted by Robert Redford at the Sundance Institute in Sundance, Utah. In addition to catering to the arts, this Artist of Cajun Cooking has developed a Creative Cajun Cooking class for the Continuing Education Program at the University of Louisiana at Lafayette.

A businessman to boot, Chef Patrick Mould co-owned a popular Lafayette eatery for a stint, and he continues to be in demand for his consulting talents. He has worked for many major corporations, such as Grand Casino, NutraSweet and the McIlhenny Company, makers of Tabasco Brand Pepper Sauce.

In 1993, Chef Patrick Mould represented McIlhenny as one of two featured chefs during the Gastronomique Fair of Dijon, the fourth largest food event in France. There, he set up a 200-seat, full-service restaurant that served over 5,000 four-course dinners during an 11-day period. The restaurant was so well received that an invitation was extended to similar fairs in Tours and Cannes, France.

Over the last decade, Chef Mould has had his fingers in a lot of pots, including his contributions to a number of cookbooks. But *Recipes From A Chef* marks his first solo effort. If you have experienced cooking Cajun cuisine, you are going to love this labor of love from Chef Mould that includes tastebud-singing recipes like Crawfish and Corn Beignets, Coconut-Battered Shrimp with Orange Marmalade Dipping Sauce, and Pumpkin Chocolate Cheesecake with Caramel Sauce.

If, on the other hand, you are a novice Cajun cook, you'll still love this cookbook. First of all, the recipes are not long or complicated. But also, characteristic of the Cajun people's unpretentious and fun-loving ways, Chef Mould's style is so relaxed and simple that you won't be intimidated preparing his exotic sounding recipes. Don't expect to fork out a lot of money to make these Nouvelle Cajun Cuisine entrées either. Around Acadiana, most at-home chefs know the practical side as well as the creative side of Chef Mould. On one of his cooking shows, for example, he looks straight into the camera as if he were talking to you across your kitchen table. He says with an encouraging smile, "Sometimes people get caught in a rut cooking the same thing over and over, and they think in order to cook a great meal they have to spend a whole lot of money. Well, we're going to prove you don't have to do that this morning!" And he does. Pushing his grocery cart down the store aisles, the cooking show host shops for the main ingredients of a three-course dinner for under $20, including the wine.

When asked how he first got interested in cooking, Chef Patrick Mould answers matter-of-factly, "All men in South Louisiana love to cook, and I was no exception." But like most great chefs, Patrick Mould has had to pay his dues. After college, he worked in the shipyards around Morgan City, Louisiana. Within a few years, the young steelworker was burnt out and decided to try his hand at cooking. He enrolled at the Lafayette Regional Vo-Tech, where he graduated in the school's first culinary occupations course in 1981.

Upon graduation from the course, Mould worked in various Lafayette restaurants before he landed a job as sous chef (right-hand man to the chef) at Café Vermilionville. Even back then, he stood out among his creative and ambitious peers. But that was almost two decades ago, back when being a chef in Acadiana was considered an unglamorous occupation and when about the

only thing a local chef sautéed was an onion for a gumbo. Chef Patrick Mould has certainly brought the South's palate a long way since then.

Patrick Mould is currently operating Louisiana Culinary Enterprises, Inc., a company that specializes in restaurant and culinary consultation. By the time this cookbook reaches bookstores, his newest venture, The Louisiana School of Cooking & Cajun Store will be in operation. Located in historic St. Martinville, birthplace of the Cajun Nation, Chef Mould will host cooking classes on Cajun & Creole cuisine. The Cajun Store will feature Louisiana products of every kind.

Notes from the Chef

Making Home Cooks Better Cooks

IT IS HARD TO BELIEVE IT HAS BEEN ALMOST 20 YEARS SINCE I DECIDED TO PURSUE a career as a professional chef. Nothing in my childhood would have indicated that I was destined to become a chef; my mother, God bless her, was not the best of cooks. Her specialties included meatball fricassée and fried Spam sandwiches. My Maw Maw on the other hand was an excellent cook. Sunday dinners at her house were always a treat. One of my happiest childhood memories was when my brother Jeff and I would go over to Maw Maw's after school and she would feed us homemade fig sweet dough pies and we would wash them down with a tall glass of ice cold milk. However, I never found myself drawn to find out how she made them, I just enjoyed the moment. Perhaps it was moments like this that led me to my love of food. Who knows?

Perhaps it was my Paw Paw, a great BBQ'er, or my uncle Shake who operates a catering business in Baton Rouge, LA. Whatever the reason, one day I just found myself gravitating toward cooking. I would do it in my spare time, when not working as an ironworker, and I came to enjoy it so much I decided in 1979 to enroll in the first Culinary Occupations course at Lafayette Regional Vo-Tech in Lafayette, LA. The cost was minimal (books and uniforms). What did I have to lose? Tom Thompson, the instructor, an ex-Navy man, was a great teacher. I attended school during the day and worked in various restaurant kitchens at night, cramming two years of study into one. I believe this was pivotal to my success to date.

My first job as a sous chef, right hand man to the chef, was under Chef Leopold Langoria at Café Vermilionville. He taught me the art of making a sauce, how to sauté a delicate piece of veal, and how to conduct myself as a professional chef. There were others along the way: Carol Boudreaux, Eula Mae, Sam & Mary, Dickie Torres, Junior Savoy—people who knew what good food was supposed to taste like. We all had a common bond: food and the preparation of it to its maximum potential. What I didn't know I searched out from other sources—chefs, cookbooks, the people of Louisiana who have been cooking great food for the past 300 years. Food in Louisiana is a passion; where people in other parts of the country eat to live, we live to eat!

I never envisioned myself cooking on television. One day I happened to be watching an episode of "Cajun Country USA" on TV-10, a show about the Cajun way of life. But one aspect of our unique culture was missing: the food. After a few phone calls I found myself in front of the camera reporting on the great restaurants and food of the Acadiana region. After a successful run on "Cajun Country," I began producing a segment for the news department called "Louisiana Cooking." I did features on Cajun culinary subjects such as the crawfish industry, andouille making and boudin making, as well as profiles of some of the great chefs in the area. Then Church Point Wholesale developed the idea of doing an in-studio cooking show that featured some of the products from their wholesale grocery line. "The Chef" airs every Monday during "Passe Partout" and "Meet Your Neighbor," two of TV-10's most popular programs. The response to the show and recipes has been incredible; for that I am grateful.

My philosophy has always been simple: use the best possible ingredients and treat them with respect. The downfall of a good recipe is the use of inferior ingredients, if you start with the best possible ingredients for a recipe, you have to work hard to mess it up. The other key ingredient is confidence. It never ceases to amaze me when people say they love to eat but can't cook. Anyone can cook with the right amount of confidence. You also have to be able to accept failure. Trial and error are the best teachers when it comes to creating a culinary masterpiece.

I am not a big fan of complicated recipes. I feel that sometimes chefs think that the more ingredients and steps you have in a recipe, the better the recipe. I think the flavor of food should be allowed to come through in a dish and not be masked with a plethora of ingredients. You will notice that most of the recipes in this book utilize minimal ingredients and that most of the ingredients can be found in any pantry in America—except of course for some of the Louisiana ingredients like tasso, andouille, crawfish and alligator. I have provided a list of companies who will be happy to provide those ingredients. Part of my goal in creating these recipes was to help home cooks become better cooks and to encourage them to inject some creativity in their day-to-day cooking. These are recipes you can use to spice up that family supper, impress your friends at a cocktail party, cook with the kids, or romance that special person in your life.

They are simply: "Recipes From A Chef."

Happy Eating!

Chef Patrick Mould C.E.C.

About the Photographer

DEBBIE FLEMING CAFFERY, WHO LIVES IN BREAUX BRIDGE, LOUISIANA, RECEIVED a degree in fine arts from the San Francisco Art Institute. Her first book, *Carry Me Home*, on the history of the sugar industry, was published in 1990. Her compelling photos have earned numerous awards, including the Governor of Louisiana's award for excellence in arts. Debbie's work is in the permanent collections of museums including the Smithsonian Institute, the Bibliotheque Nationale in Paris, and the Museum of Modern Art. In 1996 she received the prestigious Lou Stoumen Award.

Debbie's photographs are celebrated for their mystery and magic, for their spiritual shadows and darkness. Acadiana, the heart of Cajun culture in Southwest Louisiana, provided the backdrop that inspired the photos in this cookbook.

BANANA WALNUT PANCAKES WITH STRAWBERRY BUTTER

This is a great recipe to prepare with the kids and a great way to introduce them to the fun of cooking.

2 cups biscuit mix
1 cup milk
2 large eggs
1 teaspoon cinnamon
3 sliced bananas (about 3 cups)
½ cup walnut pieces
Nonstick cooking spray

STRAWBERRY BUTTER

1 stick unsalted butter
½ cup sliced strawberries
1 tablespoon powdered sugar
1 tablespoon strawberry preserves

DIRECTIONS

1. In a bowl blend together biscuit mix, milk, eggs and cinnamon until pancake batter forms. Fold in bananas and walnut pieces.
2. Heat a skillet and coat with nonstick cooking spray. Drop batter by ¼ cup into skillet. Brown pancakes on both sides and remove from pan. Repeat process until all batter has been used.
3. Top pancakes with small amount of strawberry butter and your favorite syrup.

STRAWBERRY BUTTER

Combine all ingredients in a food processor and blend until mixture is puréed and smooth. Chill butter slightly.

Yields approximately 8 to 10 pancakes.

CHOCOLATE PEANUT BUTTER PECAN COOKIES

What two ingredients are more perfectly matched than chocolate and peanut butter? The addition of pecans makes these cookies even better. My son, Ethan, loves to make them. The smell of homemade cookies baking is a childhood memory that few of us forget.

½ cup unsalted butter, softened
½ cup creamy peanut butter
½ cup granulated sugar
½ cup light brown sugar
1 egg
¾ cup flour
½ cup cocoa
2 teaspoons baking powder
½ teaspoon salt
1 cup pecan pieces

DIRECTIONS

1. Preheat oven to 350°.
2. In a bowl, place butter, peanut butter and sugars, beat until light and fluffy. Add egg and continue to beat until completely incorporated.
3. In a separate bowl sift together flour, cocoa, baking powder and salt. Add creamed butter-egg mixture and pecans; stir until a dough forms.
4. Drop cookie dough by the tablespoon onto ungreased cookie sheet pan and slightly press down with spoon, leaving edges rough.
5. Bake for 10 minutes.
6. Cool slightly before removing from cookie sheet pan or cookie may break.

Yields approximately 3 dozen cookies.

CHOCOLATE PECAN PRALINES

This popular Louisiana confection is given a little twist with the addition of chocolate.

- *3 cups granulated sugar*
- *1 stick unsalted butter*
- *1 cup evaporated milk*
- *¼ pound semisweet chocolate, coarsely chopped*
- *2 tablespoons vanilla extract*
- *1 cup pecan pieces*
- *2 cups pecan halves*

DIRECTIONS

1. Combine sugar, butter and evaporated milk in a medium saucepot over medium heat.
2. Bring to boil; lower heat to a simmer. Add chocolate and stir constantly until temperature reaches 238° on a candy thermometer.
3. Stir in vanilla, pecan pieces and pecan halves. Simmer an additional minute and remove from heat. Allow to cool for 5 minutes, stirring occasionally.
4. Drop pralines by heaping tablespoon onto parchment or wax paper and allow to sit at room temperature until pralines are set up. Store in airtight container.

Yields 20 pralines.

Mardi Gras Love

FIG NUT BREAD

This is an extremely easy and tasty bread to prepare.

1 tablespoon cooking oil
1 stick unsalted butter, softened
1½ cups fig preserves, mashed
2 eggs
½ cup milk
1 teaspoon vanilla extract
2½ cups flour
1 cup pecan pieces
½ cup chopped dates
1 teaspoon baking soda
1 teaspoon cinnamon
1 teaspoon salt

DIRECTIONS

1. Preheat oven to 350°. Grease a 9 x 5 x 3-inch loaf pan with cooking oil.
2. In a bowl cream together butter and fig preserves. Stir in eggs until completely blended.
3. Add milk and vanilla and beat until smooth.
4. Stir in remaining ingredients and pour into loaf pan.
5. Bake for 1 hour or until toothpick inserted in middle comes out clean. Cool for 5 minutes before removing from pan.
6. Allow to completely cool before slicing.

Yields 1 loaf.

HAM AND CHEESE PANCAKES

This is a great change of pace from the traditional breakfast of ham and eggs.

- 1 *cup biscuit mix*
- 1 *cup minced ham*
- 2 *eggs*
- ½ *cup milk*
- ½ *cup shredded cheddar cheese*
- ½ *cup shredded mozzarella cheese*
- 1 *teaspoon Tony Chachere's Creole Seasoning*
- ½ *teaspoon garlic powder*
- ½ *teaspoon onion powder*
- ½ *teaspoon hot sauce*
- 2 *teaspoon minced parsley*
- 1 *tablespoon minced green onions*
- *Nonstick cooking spray*
- 8 *poached eggs*
- *Cane syrup*

DIRECTIONS

1. In a bowl, combine biscuit mix, ham, eggs, milk, both cheeses, Creole Seasoning, garlic powder, onion powder, hot sauce, parsley and green onions; mix until blended.

2. Heat a nonstick skillet over medium fire and coat with nonstick cooking spray. Drop pancake batter 2 tablespoons at a time into skillet. Cook until bubbles form, flip and allow the other side to brown. Repeat process until all batter is used.

3. Place 2 pancakes on a plate and top with 2 poached eggs and drizzle liberally with cane syrup.

Yields 4 servings.

HERB BUTTERMILK DROP BISCUITS

After adding spices and herbs to your biscuits, a plain old biscuit will never taste as good. You can experiment with a variety of herbs in the recipe to develop your own version.

- 1 *cup buttermilk*
- 2 *tablespoons minced parsley*
- 1 *teaspoon dried basil leaves*
- 1 *teaspoon dried oregano leaves*
- 2½ *cups biscuit mix*
- 1 *teaspoon Tony Chachere's Creole Seasoning*
- 1 *teaspoon granulated onion*
- ½ *teaspoon granulated garlic*
- ¼ *teaspoon white pepper*
- ¼ *teaspoon salt*
- ¼ *cup shredded Parmesan cheese*
- 3 *tablespoons unsalted butter, melted*

DIRECTIONS

1. Preheat oven to 400°. In a bowl, stir together buttermilk, parsley, basil and oregano. Set aside.
2. In a another mixing bowl, blend together biscuit mix, Creole seasoning, onion, garlic, white pepper, salt and Parmesan cheese.
3. Add buttermilk-herb mixture to biscuit mixture and stir until soft dough is formed.
4. Drop by the heaping spoonful onto an ungreased baking sheet pan. Brush with melted butter and bake for 15 minutes or until golden brown.

Yields 18 biscuits.

NEW ORLEANS MUFFULETTA

This is one of the heartiest sandwiches known to man. If you are unable to find muffuletta bread, ask your local baker to prepare you a special batch.

OLIVE SALAD DRESSING

- 2 *(10-ounce) jars Italian Olive Salad, drained*
- 1 *(6-ounce) jar pitted black olives, drained and minced*
- ½ *cup extra virgin olive oil*
- 2 *tablespoons balsamic vinegar*
- 2 *teaspoons dried basil leaves*
- 2 *teaspoons dried oregano leaves*
- 1 *teaspoon dried thyme leaves*
- 1 *teaspoon Tony Chachere's Creole Seasoning*
- 1 *teaspoon hot sauce*

MUFFULETTA

- 1 *muffuletta bread, split*
- 12 *tablespoons olive salad dressing*
- 4 *slices* ***Bryan Cotto Salami***
- 4 *slices* ***Bryan Deli Classic Ham***
- 4 *slices* ***Bryan Deli Classic Smoked Turkey***
- 2 *slices mozzarella cheese, halved*
- 2 *slices provolone cheese, halved*

DIRECTIONS

OLIVE SALAD

In a bowl stir together all ingredients until blended. Any leftover olive salad can be stored in an airtight container in the refrigerator.

MUFFULETTA

1. Lay muffuletta bread cut side up and spread each half of bread with a coating of 6 tablespoons of olive salad dressing.
2. Alternate **Bryan sliced meats** and cheese slices on one half of bread until stacked nicely. Top with other half of muffuletta bread. Press down on muffuletta.
3. Heat skillet and spray with nonstick spray. Cook on both sides until bread is toasted. Cut in half, then into quarters and serve.
4. If cheese is not melted, microwave for a minute or place in 350° oven for several minutes.

Yields 2 or 4 servings.

Jamming at Whiskey River Landing with Christine Balfa and Sheryl Cormier

PAIN PERDU

"Pain Perdu" is French for lost bread because it is best made with stale French bread. By soaking it in the egg and sugared milk mixture it is transformed into a breakfast worthy of a king.

4 eggs
½ cup milk
¼ cup sugar
1 tablespoon vanilla extract
1 teaspoon cinnamon
½ teaspoon nutmeg
8 (1-inch) slices of French bread, cut on a diagonal
2 tablespoons unsalted butter
Cane syrup
Confectioners' sugar

DIRECTIONS

1. In a bowl, combine eggs, milk, sugar, vanilla, cinnamon, and nutmeg; whip until blended.
2. Place bread in egg batter, four slices at a time, and soak for 1 minute on each side to absorb batter. Do not allow bread to soak too long or it will become soggy and hard to handle.
3. Heat 1 tablespoon butter in frying pan over medium heat. Fry bread on both sides until completely browned. Repeat process with remaining 4 slices of French bread.
4. Place two slices of pain perdu on a plate and drizzle with liberal amount of cane syrup and sprinkle with lots of confectioners' sugar.

Yields 4 servings.

PASTA MACQUE CHOUX

This pasta dish is made with a wealth of ingredients and is very rich. It can be served by itself or utilized as a base for the Paneed Breast of Chicken recipe on page 84.

2 tablespoons unsalted butter
1/3 cup chopped smoked tasso
1/3 cup sliced andouille or smoked sausage
1 cup cooked corn
1/2 cup chopped onion
1/4 cup chopped green bell pepper
1/4 cup chopped celery
1 teaspoon minced garlic
3 cups heavy whipping cream
3 tablespoons grated Parmesan cheese
1 1/2 teaspoons dried basil leaves
1 teaspoon Tony Chachere's Creole Seasoning
1 teaspoon hot sauce
3/4 teaspoon dried oregano leaves
1/2 teaspoon dried thyme leaves
3 tablespoons chopped green onions
1 tablespoon chopped parsley
4 cups of your favorite cooked pasta

DIRECTIONS

1. In a large skillet, heat butter over medium heat.
2. Add tasso and sausage; brown lightly.
3. Add corn, onion, bell pepper, celery and garlic; cook for 2 minutes.
4. Add heavy whipping cream, Parmesan, basil, Creole seasoning, hot sauce, oregano and thyme. Simmer for 3 minutes until cream reduces and starts to thicken slightly. Be careful not to reduce too much because cream could curdle.
5. Stir in green onions and parsley.
6. Serve over heated cooked pasta.

Yields 4 servings.

PECAN PRALINES

Pralines are the most popular of Louisiana confections; no holiday meal would be complete without pralines.

1 ½ cups granulated sugar
1 ½ cups brown sugar
1 cup evaporated milk
2 sticks unsalted butter
3 tablespoons cane syrup
3 cups pecan pieces
2 tablespoons vanilla

DIRECTIONS

1. In a saucepot, combine granulated sugar, brown sugar, evaporated milk, butter and cane syrup.
2. Bring to boil, then lower heat and simmer, stirring constantly, until temperature reaches 238° on candy thermometer.
3. Stir in pecan pieces and vanilla. Simmer an additional minute and remove from heat. Allow to cool for 5 minutes, stirring occasionally.
4. Drop pralines by heaping tablespoon onto parchment paper or wax paper and allow to sit at room temperature until pralines are set up. Store in an airtight container.

Yields 20 pralines.

CREOLE SAUSAGE BREAD LOAVES

This is my version of a dish first introduced to me by Creole cook Merlyene Herbert. She and her husband, Sam, operate one of my favorite places to eat, the Creole Lunch House in Lafayette, La. When I have guests from out of town, I always take them there; it's a great place to introduce someone to Louisiana Soul food.

- 1 *pound* ***Bryan Pampered Pork Sausage***
- ½ *pound* ***Bryan Smoked Sausage,*** *quartered and sliced*
- 1 *cup chopped smoked tasso*
- 1 *cup diced onion*
- ½ *cup diced bell pepper*
- ½ *cup diced celery*
- 1 *teaspoon hot sauce*
- ½ *teaspoon Tony Chachere's Creole Seasoning*
- 4 *tablespoons flour*
- 1½ *cups beef broth*
- ¼ *cup shredded cheddar cheese*
- 2 *tablespoons minced green onions*
- 2 *(1-pound) frozen bread dough loaves, thawed*

DIRECTIONS

1. Preheat oven to 325°.
2. Heat saucepot over medium heat and add **Bryan Pampered Pork Sausage** and **Bryan Smoked Sausage.** Cook until sausage starts to brown. Drain on paper towels.
3. Add tasso, onion, bell pepper, celery, hot sauce and Creole seasoning. Cook for 10 minutes. Add flour and continue to cook for 5 minutes, scraping bottom of pan with spoon.
4. Add beef broth and cook for an additional 10 minutes until meat mixture is thick. Add cheese and green onions and cook for 5 minutes.
5. Pour into bowl and allow to cool in refrigerator.
6. Place defrosted bread dough balls on floured surface and roll out to ½-inch thick rectangles (about 4 x 8 inches).
7. Divide filling into four parts and spread two strips of filling lengthwise onto each rolled-out bread dough loaf leaving space between the two strips.
8. Fold about one inch of each end toward center. Then roll dough, using the long side, into a cylinder like a cigar. Seal ends with a dab of water.
9. Bake for 40 minutes.
10. Allow to cool slightly then cut loaves into slices about 2 inches thick.

Yields 2 loaves.

SWEET AND SPICY PECANS

These nuts are great to eat while watching your favorite sporting event. They are also a nice, crunchy addition to your favorite salad.

3 tablespoons unsalted butter
1 tablespoon minced garlic
2 tablespoons brown sugar
2 tablespoons sugar
2 tablespoons hot sauce
1 tablespoon chili powder
½ teaspoon salt
3 cups pecan halves

DIRECTIONS

1. Preheat oven to 200°.
2. In skillet heat butter. Add garlic and cook for 1 minute.
3. Add remaining ingredients except pecans and cook 2 minutes.
4. Place pecans and seasoned butter in a bowl and toss until pecans are coated.
5. Place on cookie sheet and bake for 30 minutes, stirring every 10 minutes.
5. Allow to cool before eating.

Yields 3 cups.

Nathan Williams Jr. and Mark Willams on washboards

Recipe Sponsors

Louisiana Product Guide

CHEF PATRICK MOULD'S LOUISIANA SCHOOL OF COOKING & CAJUN STORE
112 South Main Street
St. Martinville, LA 70582
(337) 394-1710
e-mail-patrick.mould@gte.net
www.lacooks.com
Cooking classes, Louisiana food products, Cajun products and Culinary Adventures

DEBBIE FLEMING CAFFERY
105 Washington St.
Breaux Bridge LA 70517
(337) 332-6254
caffery@mindspring.com
Fine art photography & prints

NEW ORLEANS FISH HOUSE
921 South Dupre Street
New Orleans, LA 70125
(800) 839-3474
Shrimp, crawfish, crabs, oysters, fish, soft shell crabs

TONY CHACHERE'S CREOLE FOODS
519 North Lombard
Opelousas, LA 70570
(800) 551-9066
www.cajunspice.com
Creole seasonings

SAVOIE'S ROUX
1742 Hwy. 742
Opelousas, LA 70570
(337) 942-7241
Dark & light roux, oilless roux

FALCON RICE MILL
P.O. Box 771
Crowley, LA 70527
(337) 783-3825
Louisiana Rice-Toro Rice, Cajun Country Popcorn Rice

JACK MILLER'S BARBECUE SAUCE
P.O. Box 57
Ville Platte, LA 70586
(800) 646-1541
www.jackmillers.com
Louisiana style barbecue sauce

POCHE'S MARKET RESTAURANT & SMOKEHOUSE
3015-A Main Hwy.
Breaux Bridge, LA 70517
(337) 332-2108
(800) 3-Poches
Smoked tasso & andouille

EL SIDO'S NIGHT CLUB
1523 N. St. Antoine St.
Lafayette, LA 70501
(337) 237-1959
Zydeco & Blues music & Creole cooking

NATHAN WILLIAMS & THE ZYDECO CHA CHAS
c/o Concerted Efforts
Attn: Paul Kahn
(617) 969-0810
Manager: Sidney Williams
(337) 235-0647
468@bellsouth.net
The Best in Zydeco

ANGELLE'S WHISKEY RIVER LANDING
1365 Henderson Levee Road
Henderson, La. 70517
Contact: Terry Angelle
(337) 228-8567
www.angelleswhiskeyriver.com
Cajun Dancehall & Boat Landing

BALFA TOUJOURS
Under The Hat Productions
1121-B Bluebonnet Lane
Austin, Texas 78704
Contact: Cash Edwards
(512) 447-0544
cashuthp@earthlink.net
Traditional Cajun Music

SHERYL CORMIER & CAJUN SOUND
4917 NW Evangeline Thruway
Carencro, LA 70520
(337) 896-0652
Traditional Cajun Music

Chef's Tips

- I use dried herbs in all the recipes because this is usually what most people have in their cupboard. If you have access to fresh herbs, by all means use them. Always double the amount needed for the recipes when using fresh herbs.
- Canned broths are used for all the recipes. If you are using made-from-scratch stocks you will have to adjust the salt to taste. Always use low-sodium low-fat canned broths when possible.
- I prefer to use unsalted butter in all the recipes, especially when baking.
- When making a salad dressing I prefer to use extra-virgin olive oil for its intense flavor. When making marinades or sautéing it is not necessary to use extra-virgin olive oil.
- If you are using wild alligator meat for the Alligator Smoked Andouille Sauce Piquante on page 44, you will have to increase the cooking time in step 1 in order to tenderize the wild alligator. If you are using farm-raised alligator, the cooking times will remain the same; either way, it's going to taste like chicken.
- Tasso is pork that has been marinated in a spicy brine prior to it being smoked. If you can't find tasso where you live you may substitute a spicy smoked ham.
- Andouille is a lean, chunky smoked sausage found in Louisiana.
- When stirring butter into a sauce make sure it is chilled and the fire is as low as it will go while stirring in the butter. This will insure that the sauce will hold together and thicken.

A

B

C

P

T

Published by Louisiana Culinary Enterprises, Inc.
Lafayette, LA (337) 983-0896

ISBN: 0-9673971-0-3

First Printing

December 1999

5,000 copies

Photography by:

Debbie Fleming Caffery

For additional copies see order form on next page.

RECIPES FROM A CHEF

Send check or money order to:
Louisiana Culinary Enterprises, Inc.
P.O. Box 90331
Lafayette, LA 70509

Credit Card Orders-Phone (337) 394-1710
Visa, Mastercard, American Express and Discover accepted

Please send me ____ copy(ies) of **RECIPES FROM A CHEF**

$21.95 per copy ________________
$ 4.95 Shipping & Handling ________________
(Louisiana residents, please add 7.5% sales tax.) ________________

Name ____

Address ____

City ______________________ State __________ Zip ____________

If you wish to have it personalized include the person's name to whom the book is being sent along with their address, and Chef Pat will personally autograph the cookbook.

- -

RECIPES FROM A CHEF

Send check or money order to:
Louisiana Culinary Enterprises, Inc.
P.O. Box 90331
Lafayette, LA 70509

Credit Card Orders-Phone (337) 394-1710
Visa, Mastercard, American Express and Discover accepted

Please send me ____ copy(ies) of **RECIPES FROM A CHEF**

$21.95 per copy ________________
$ 4.95 Shipping & Handling ________________
(Louisiana residents, please add 7.5% sales tax.) ________________

Name ____

Address ____

City ______________________ State __________ Zip ____________

If you wish to have it personalized include the person's name to whom the book is being sent along with their address, and Chef Pat will personally autograph the cookbook.

RECIPES FROM A CHEF

Send check or money order to:
Louisiana Culinary Enterprises, Inc.
P.O. Box 90331
Lafayette, LA 70509

Credit Card Orders-Phone (337) 394-1710
Visa, Mastercard, American Express and Discover accepted

Please send me _____ copy(ies) of **RECIPES FROM A CHEF**

$21.95 per copy ________________

$ 4.95 Shipping & Handling ________________

(Louisiana residents, please add 7.5% sales tax.) ________________

Name ____

Address ____

City ______________________ State __________ Zip ______________

If you wish to have it personalized include the person's name to whom the book is being sent along with their address, and Chef Pat will personally autograph the cookbook.

RECIPES FROM A CHEF

Send check or money order to:
Louisiana Culinary Enterprises, Inc.
P.O. Box 90331
Lafayette, LA 70509

Credit Card Orders-Phone (337) 394-1710
Visa, Mastercard, American Express and Discover accepted

Please send me _____ copy(ies) of **RECIPES FROM A CHEF**

$21.95 per copy ________________

$ 4.95 Shipping & Handling ________________

(Louisiana residents, please add 7.5% sales tax.) ________________

Name ____

Address ____

City ______________________ State __________ Zip ______________

If you wish to have it personalized include the person's name to whom the book is being sent along with their address, and Chef Pat will personally autograph the cookbook.